MicroRNAs as Tools in Biopharmaceutical Production

Niall Barron
Editor

MicroRNAs as Tools in Biopharmaceutical Production

Springer

Editor
Niall Barron
National Institute for Cellular Biotechnology
Dublin City University
Glasnevin, Dublin
Ireland

ISBN 978-94-007-9526-6 ISBN 978-94-007-5128-6 (eBook)
DOI 10.1007/978-94-007-5128-6
Springer Dordrecht Heidelberg London New York

Softcover reprint of the hardcover 1st edition 2012

Printed on acid-free paper.

Springer is part of Springer Science+Business Media (www.springer.com)

Preface

Since their discovery in 1993 by the Ambros lab there has been a phenomenal increase in research into microRNAs. There are now thousands of peer-reviewed publications describing all aspects of their biogenesis and function, and indeed the resulting diseases when they dysfunction. Yet considerable gaps remain in our understanding of how these tiny regulators of gene expression are borne, live and die. What is overwhelmingly obvious is that they play a critical role in many if not all important cellular processes. Their regulatory niche within the cell lies at the interface between the mRNA transcript and its translation into a functional protein providing the cell or organism with yet another mechanism to fine-tune gene expression. Like transcription factors, individual miRNAs also possess the ability to interact with and influence the expression of tens to hundreds of target genes providing a mechanism to rapidly influence entire cellular pathways or processes. Importantly, this intervention can be deployed quickly (and at a cheaper metabolic cost) due to the absence of a translational step—they are functional without the need to build a protein.

It is these two properties in particular that make them of potential interest to those involved in the production of recombinant proteins by mammalian cells. The desire to improve efficiencies in the biosynthesis of Biopharmaceutical products, including monoclonal antibodies, blood factors, fusion proteins and other exotic polypeptides continues to motivate research into improving culture medium, bioreactor design, vector design and process control. In the last decade there have been a number of research groups in both industry and academia that have focused on what many believe to be the source of the next major improvements in process efficiencies—genetically engineering the producer (usually CHO) cell. The goal is typically to bestow improved phenotypic characteristics, such as rapid proliferation to high density, prolonged viability in the challenging environmental conditions in late stage culture, reduced reliance on particular substrates or reduced sensitivity to waste metabolites, and high specific secretion rates of an active, intact protein product with the most desirable post-translational modifications. The search for key molecules responsible for these cellular properties has resulted in numerous publications describing genes and proteins whose expression changes in CHO cells under different culture conditions or in cells with different phenotypic characteristics. There have been several examples of exogenous dysregulation of particular genes leading to improvement in these traits though it is less clear how many of these have made their way into actual industrial processes. One of the challenges associated with this approach is the fact that complex traits, such as proliferation, are necessarily controlled by multiple genetic interactions meaning that frequently, engineering the expression of a single gene may not be sufficient to influence the phenotype. On the other hand, exogenous overexpression of multiple protein-coding genes places a metabolic burden on cells whose main role is to generate large quantities of the product, hence causing intra-cellular competition for biosynthetic resources. miRNAs, therefore, could provide an alternative means to influence the expression of multiple genes simultaneously in order to modify the pathways and processes underpinning particular phenotypes without competing with product protein synthesis.

This volume aims to provide those interested in the potential utility of miRNAs in the bioprocessing field with a succinct overview of what is known about these fascinating molecules—their biogenesis, mode of action, known functions in the cell, how they can be detected, measured and modified—with a particular focus on relevance to recombinant protein production in CHO cells.

Dublin
2012

Niall Barron

Contents

List of Contributors

Niall Barron, PhD Dublin City University, Ireland

Michael Betenbaugh, PhD Johns Hopkins University, MD

Nicole Borth, PhD Austrian Center of Industrial Biotechnology, Vienna

Colin Clarke, PhD Dublin City University, Ireland

Stephen L. Clarke, PhD Dept. of Nutritional Sciences, Oklahoma State University

Martin Clynes, PhD Dublin City University, Ireland

McKale R. Davis, PhD Dept. of Nutritional Science, Oklahoma State University

Aliaksandr Druz, PhD NIDDK, NIH, Bethesda, MD

Mark Gallagher, BSc Dublin City University, Ireland

Johannes Grillari, PhD BOKU, University of Vienna

Matthias Hackl, PhD BOKU, University of Vienna

Michael Henry, MSc Dublin City University, Ireland

Lyne Jossé, PhD University of Kent, UK

Carol Kreader, PhD SAFC/Sigma-Aldrich, Saint Louis

Nan Lin, PhD SAFC/Sigma-Aldrich, Saint Louis

Paula Meleady, PhD Dublin City University, Ireland

Joseph Shiloach, PhD NIDDK, NIH, Bethesda, MD

C. Mark Smales, PhD University of Kent, UK

Raymond L. Stallings, PhD Cancer Genetics, Royal College of Surgeons in Ireland

Ramanjulu Sunkar, PhD Dept. of Biochem & Mol.Biol., Oklahoma State University

Lin Zhang, PhD Pfizer Inc, Andover, MA

Chapter 1
A General Introduction to MicroRNAs, Their Investigation and Exploitation in CHO Cell Lines

Lyne Jossé, Lin Zhang and C. Mark Smales

Abstract MicroRNAs are small (20–23 nucleotides) non-coding RNA molecules that mainly act as negative regulators of gene expression by binding to target mRNAs in their 3′UTR. In this chapter, we provide a brief account of the *in silico* and experimental tools available for researchers working in protein engineering in eukaryotic cell factories. In considering the wide influence of microRNAs, we place a special emphasis on the cellular effects associated with the manipulation of microRNA and high yield of protein production with reference to use in Chinese hamster ovary (CHO) expression systems.

Keywords MicroRNA biogenesis · MicroRNA editing · MicroRNA binding specificity · MicroRNA stability · MicroRNA abundance

1.1 Introduction

MicroRNAs are central to many cellular processes and can regulate a broad range of targets. This versatility potentially opens many new windows of opportunity for the biopharmaceutical industry in terms of controlling gene expression and cell phenotype in industrial CHO cell lines, particularly as cell engineering strategies which target multiple genes are increasingly accepted as the most likely to impact upon desired phenotypes. To date the majority of studies on these molecules have been undertaken with respect to cancer therapy and much of this knowledge can be directly related to industrial phenotypes (e.g. those that control growth). Whilst still a relatively poorly investigated area of gene regulation, the field has benefited from the development of deep sequencing and RNA sequencing technologies coupled with comprehensive database platforms which are regularly updated. In addition to the characterisation of microRNA dynamics, the current effort to annotate the CHO genome will most certainly accelerate the progression of microRNA investigations

L. Jossé (✉) · C. M. Smales (✉)
Centre for Molecular Processing and School of Biosciences, University of Kent,
CT2 7NJ Canterbury, Kent, UK
e-mail: c.m.smales@kent.ac.uk

L. Zhang
Pfizer Inc, 1 Burtt Road, Andover, MA 01810, USA

N. Barron (ed.), *MicroRNAs as Tools in Biopharmaceutical Production*,
DOI 10.1007/978-94-007-5128-6_1, © Springer Science+Business Media Dordrecht 2012

and cell line development strategies. However, whilst the potential to manipulate microRNAs for biotechnological benefit holds much promise, it has a number of pitfalls that are partly due to the fundamental principles of miRNA:mRNA target recognition and the difficulty in predicting to what degree the titration/manipulation of a particular microRNA will impinge on its one or many mRNA targets. In this chapter, we provide a brief account of the *in silico* (e.g. target prediction software) and experimental (e.g. reporter genes and sponge vectors) tools available for researchers working in microRNAs and the manipulation of mammalian cell factories. In considering the wide influence of microRNAs, we place a special emphasis on the cellular effects associated with the manipulation of microRNA and high yield of protein production with reference to use in Chinese hamster ovary (CHO) expression systems.

1.2 Biogenesis and Evolution

MicroRNAs were first described in *C. elegans* as an antisense RNA regulating the level of LIN-14 protein (Lee et al. 1993). The number of small RNAs described has since exploded. As an example the online repository miRBase (http://mirbase.org/) release 18.0, November 2011 contains 18,226 entries representing hairpin precursor miRNAs, expressing 21,643 mature miRNA products in 168 species. These include 1,527 in human, 741 in mouse, 408 in rat and surprisingly only two have been deposited for the hamster.

The maturation pathways of microRNAs are now well defined and the field is extensively reviewed elsewhere (Kim et al. 2009; Lau and MacRae 2009; Westholm and Lai 2011; Winter et al. 2009). In summary, microRNA biogenesis is initiated by the transcription (usually by Pol II) of hairpin-containing RNAs, which are known as pri-miRNAs or mIRtrons in the case of intronic miRNAs. In the first scenario, the primary transcripts are cleaved into 55–70 nt hairpins, called pre-miRNAs, by the nuclear RNase III multicomplex Drosha and its dsRBD partner DGCR8 (Han et al. 2004; Lee et al. 2003) and is subsequently exported to the cytoplasm. For intron-derived microRNAs, the consensus is that Drosha is substituted by splicing although cleavage of the miRNA hairpin can occur prior to the splicing of mRNA (Kim and Kim 2007). Subsequently, miRtrons, which reside in the cytoplasm, undergo a process known as debranching before adopting a pre-miRNA-like structure (Ruby et al. 2007). Both pre-miRs are then further shortened by Dicer RNase III to yield a 22 nt miRNA known as the miRNA/miRNA* duplex, which exhibit characteristic 3′overhangs at either end (Hutvagner et al. 2001; Kim et al. 2009). The strand that accumulates to a higher level is referred to as the "mature" miRNA, while its less abundant partner is referred to as the miRNA* or "star" strand. The abundant strand is thought to be preferentially incorporated into an Argonaute (AGO) protein, which acts as the core of an effector complex that is routed by the small RNA to targets and the remaining strand is degraded.

The enzymes involved in the biogenesis of microRNAs are broadly conserved and many of the bilaterian animal mature miRNAs are also phylogenetically conserved;

55 % of *C. elegans* miRNAs are related to human miRNAs and ~55 % *Drosophila* miRNAs are related to human (Ibanez-Ventoso et al. 2008) indicating that miRNAs have had important roles throughout animal evolution and implying that miRNAs are conserved across several species and tightly regulate their target. Indeed, target genes are more efficiently repressed by conserved miRNAs as opposed to less conserved microRNAs (Huang and Gu 2011).

As described above, microRNAs can be found in both intergenic and intragenic regions, sometimes in clusters. A detailed analysis of 60 microRNAs by Lagos-Quintana et al (2003) revealed that 33 of these 60 resided in intergenic regions, 7 in sense orientation in non-coding genes/regions, and 13 in sense/7 in antisense orientation within introns. The apparent discrepancy of microRNA loci of expression raised at the time a question as to which promoter elements and RNA polymerases controlled the transcription of microRNAs. It has since been ascertained that the majority of pri-miRNAs are transcribed by RNA polymerase II (Cai et al. 2004), but a fraction of these, notably residing upstream of tRNA sequences, Alu sequences or mammalian-wide interspersed repeat (MWIR) sequences, are transcribed by RNA polymerase III (Borchert et al. 2006).

1.3 Informative Tools to Predict Targets of MicroRNAs

1.3.1 Target Binding and Specificity

Target genes can be paired to more than one miRNA, likewise a single miRNA could in theory target hundreds of unrelated messages and as a result it is thought that almost a third of genes are potentially regulated by microRNAs (Lewis et al. 2005). This is essentially because in mammalian cells the target recognition is mediated via imperfect Watson-Crick base pairing to the mRNA *i.e.* only 6–8 nucleotides (called the seed region) near the 5′ end of the miRNA show a perfect match (as opposed to plants where there are few or no mismatches). Notably, the 3′ region may supplement seed matches or compensate for mismatches (Bartel 2009). Whilst the sequence differs primarily at the 3′ and 5′-end of the molecule, the biological significance of these observations is unclear. Kiryu et al also suggested that miRNAs have positive correlations with accessibilities in broad regions downstream of their putative binding sites, which could indicate that downstream regions of the target sites are bound by other proteins, assisting miRNA binding (Kiryu et al. 2011).

To date, only a fraction of miRNAs have been assigned a biological function and therefore recent efforts have focused on developing prediction software to identify additional miRNA targets in a specific host in order to further annotate pathways of regulation. Current prediction tools differ in their approach and performance and sometimes exhibit poor overlap. However, TargetScan (Friedman et al. 2009), PicTar (Krek et al. 2005), Miranda (Betel et al. 2008), PITA Top (Kertesz et al. 2007), mIRGator (Nam et al. 2008) and the combined use of these software is a good starting point to identify potential miRNA:mRNA interactions (for review, see Li et al. 2010, Min and Yoon 2010). To date such approaches have been limited in CHO due to the lack of a publicly available annotated genome, however with the recent publication of the CHO genome such limitations should soon be removed.

1.3.2 MicroRNA Networks

The hundreds of targets predicted to be influenced by a single microRNA suggest a complex interplay between all genes involved. MicroRNAs are known to play an important role in development, proliferation, differentiation, and apoptosis (Wijnhoven et al. 2007; Bueno et al. 2008). Controlling all of these aspects of cell behaviour is crucial to the development of robust industrial cell factories. A better understanding of the miRNA-mediated network of genes and cellular pathways is essential if one is to manipulate miRNA in an informed manner with the aim of improving the cell "biology" and enhance its bioprocessing capacity. A few bioinformatics tools dedicated to network analysis have recently emerged that may assist this approach. For, example DIANA-mirPatch identifies molecular pathways potentially altered by the expression of one or several microRNAs by looking at KEGG pathway enrichment. Unfortunately, up to now these web-tools remain limited to a set of microRNAs and exclude the CHO genome. Recently, Hsu et al. (2008) studied the human microRNA-regulated protein-protein interaction (PPI) network by utilizing the Human Protein Reference Database (HPRD) and the miRNA target prediction program TargetScan. They found that an individual miRNA often targets the core gene of the PPI network. Despite advances in such *in silico* prediction software, the construction of a microRNA network must be supported by sufficient experimental data. With regard to this, of particular use for identifying microRNAs that may play a role in an industrial bioprocessing sense, has been the development of (i) microRNA arrays that allow the monitoring of microRNA expression of a library of microRNAs associated with a particular phenotype of interest, (ii) RNA sequencing, (iii) qRT-PCR assays for the accurate measurement of microRNA amounts in a given sample, (iv) over-expression and sponge vectors to increase and reduce microRNA levels respectively. These tools, alongside the in silico prediction tools, now provide the investigator with a suite of tools to predict, screen for, and validate the influence of microRNA levels on cell phenotype. Such approaches can now be aligned with transcriptomic and proteomic investigations that can identify the target mRNAs and subsequent proteins that microRNA manipulation ultimately influences.

1.4 Parameters that Govern Repression

1.4.1 MicroRNA Stability and Abundance

Aside from the prediction of the microRNA binding sites, the 3′UTR local features, the actual number and respective abundance of the potential targets must be considered in any experimental design as they all impact on the microRNA binding. A study by Arvey et al. examined several predictors of downregulation (A + U content, 3′UTR length and the expression level of individual transcripts) and concluded that the target binding by a miRNA was mostly influenced by the microRNA over mRNA abundance ratio (Arvey et al. 2011). This ratio is also linked to the number of predicted targets. The "dilution factor" therefore means that endogenous or plasmid-borne miRNAs ultimately may have a different influence on their specific targets depending on the host, target expression level, and culture conditions and as such these parameters must be considered when manipulating microRNAs. As stated above, the length of the target 3′UTR can influence mRNA regulation by microRNAs and Hinske et al observed that for microRNAs that reside in intronic regions, the 3′UTR of host genes predicted to be targeted by their own microRNA (20 %) are significantly longer and hold more adenylate/uridylate-rich elements (AREs) (Hinske et al. 2010). Qiu et al (2010) also examined how miRNAs evolved with respect to their target genes and transcription factors (TFs) and showed that microRNAs have different evolutionary rules than TFs and protein-coding genes.

1.4.2 RNA Editing

A-to-I RNA editing, which causes the conversion of single adenosines into inosines, is mediated by adenosine deaminases acting on RNA (ADARs) (Yang et al. 2006), a family of enzymes that recognise double-stranded RNA molecules. The process can interfere with the efficiency of microRNA maturation and thus with the miRNA homeostasis within a cell (Godfried Sie and Kuchka 2011; Gommans 2011). It appears to affect 10 % of primiRNAs downstream processing by interfering with Drosha-mediated cleavage. Interestingly, there are very few examples of mature microRNAs influenced by RNA editing although this may be an artefact of the small RNA sequences. Strikingly, RNA editing is also implicated in *de novo* generation of miRNA binding sites in Alu sequences (Borchert et al. 2009).

1.4.3 Control of MicroRNA Expression

How miRNA levels are controlled is less well described and understood. The promoter of protein-coding genes and microRNA are known to share common regulatory elements (Lee et al. 2007). Putative binding sites for transcription factors of conserved genomic regions can be found on the UCSC Genome Browser, JASPAR (http://jaspar.binf.ku.dk/) and TESS (http://www.cbil.upenn.edu/cgi-bin/tess/tess). It remains to be determined whether a set of mRNAs regulated by an individual miRNA simply reflects a random set of functionally-independent genes or not. Interestingly, Cui and colleagues (2007) concluded that genes with more transcription factor - binding sites have a higher probability of being targeted by miRNAs and have more miRNA-binding sites, suggesting that transcription factors and microRNAs might act coordinately.

1.5 Cellular Effects of MicroRNA Manipulation

1.5.1 Signalling Pathways

Proteins involved in signalling cascades are often more likely to influence the cell phenotype than other effectors. Consequently, even a subtle fluctuation in their levels via miRNA binding could have drastic consequence. Saj and Lai examined the relationship between microRNA–mediated regulation and known signalling pathways and reported evidence of the role of TGF-β/BMP and Ras/MAPK in modulating microRNA biogenesis (Saj and Lai 2011). The p53 pathway is also known to module miRNA biogenesis or expression upon DNA damage (for review see (Suzuki and Miyazono 2010)). On the other hand, some miRNAs have been shown to regulate p53 (Park et al. 2009) and thus the relationship between regulation of microRNAs and the regulation of targets by microRNAs is still a relatively unexplored area of microRNA biology.

1.5.2 MicroRNA and Stress

The principles that govern target capture such as levels of mRNAs and the number of mRNA targets remain true under conditions of stress (Stern-Ginossar et al. 2008; Kroll et al. 2010). As exemplified in the above section, miRNA- mediated stress responses, for example via the p53 pathway, can involve intricate levels of regulation. In the context of recombinant protein production, an abnormal stress is placed on the host cells, often activating stress pathways such as those involved in protein folding and the folding machinery (UPR), secretion, energy metabolism, lipid metabolism and membrane biogenesis in protein secreting cells. Yang et al (2011) showed that miR-122 negatively regulates UPR chaperones via the CDK4-PSMD10 pathway. A similar line of evidence, showed that miR-30c-2* was up-regulated during UPR activation, *concomitant with* Xbp1 and could bind the 3′UTR of Xbp1 (Byrd and Brewer 2011). Another hallmark of cell stress is the re-localization of protein effectors in specialized cell compartments such as Russell bodies (aggregated proteins), and stress granules (RNAs). Following cellular insults, stress-sensing kinases induce the formation of stress granules-via eIF2alpha phosphorylation and miRNAs which normally reside in the cytoplasm can be shuttled to this self-organized compartment.

1.5.3 MicroRNA and Culture Conditions

In biotechnology, the growth rate, the lifespan and the productivity of the host cells are major determinants of recombinant protein production. As a result, cell engineering strategies such as the manipulation of cell cycle progression and apoptosis have resulted in improved cell lines that can grow to very high densities in suspension cell cultures (for review see Fussenegger and Bailey 1998; Kim et al. 2011). To improve the foreign protein productivity, operators can rely on biphasic cultures whereby cells are encouraged to proliferate at 37 °C until the culture reaches the critical biomass and then the temperature is lowered (28–33 °C) to promote a longer and more viable stationary/production phase (Trummer et al. 2006; Yoon et al. 2003a, b).

The link of microRNA expression and control with cancer (Calin et al. 2002) and subsequently the regulation of apoptosis (Brennecke et al. 2003) prompted several microRNA hybridization array analyses of bioprocessing cell lines, with the aim to identify growth-phase dependent microRNA markers. For example, Gammel et al took advantage of the fact that microRNAs are highly conserved and used non-hamster arrays on their CHO model (Gammell et al. 2007). Using this approach, they showed that miR-21 and miR-24 were up-regulated in CHO-K1 growth-arrested cell lines, induced either by temperature shift or during normal batch culture. An ensuing publication into microRNA expression in CHO-K1 cells showed that lowering the temperature from 37 °C to 31 °C led to increased levels for: miR-219, miR-518d, miR-126, miR-30e, miR-489 and miR-345, when measured after 24 h, while miR-7, miR-320, miR-101 and miR-199 were down-regulated (Barron et al. 2011). Similar lines of investigation were carried out by Koh et al in human embryonic kidney 293 (HEK293) cells who examined the microRNA expression profiles in batch cultures at different growth phases (Koh et al. 2009). They reported that 13 out of 14 microRNAs, including for example hsa-let-7b and hsa-miR-16, were up-regulated in the transition from exponential to stationary phase indicative of translation slowing down for a large number of proteins.

Whilst such studies have their merits, they also emphasize that the microRNA signature greatly fluctuates with the culturing conditions as well as the type of producing cells. Interestingly, proliferating cells appear to express mRNAs with shortened 3′UTRs and UTRs that contain less miRNA target sites (Sandberg et al. 2008), indicating that some mRNA isoforms may evade post-transcriptional regulation in a growth-phase dependent manner. The cell density has also been reported to impact on microRNA biogenesis (Hwang et al. 2009) with confluent cells showing an increase in Drosha processing. These observations show that careful monitoring of all aspects of cell culturing are crucial for the accurate analysis of miRNA expression and suggest that many miRNAs may only have a short time span to fulfil their action.

1.6 Validation of MicroRNAs Target Predictions in Bioprocessing Applications

Chinese hamster ovary (CHO) cell lines are presently the gold standard mammalian expression system for the production of recombinant proteins in the biopharmaceutical industry. The majority of mammalian expressed and approved biotechnology-derived biopharmaceuticals are produced in CHO cells and it is expected that this will remain the case for the foreseeable future. Despite being at the forefront of the bioprocessing industry, up until recently, a significant disadvantage and obstacle to undertaking cellular function work with CHO cells was the lack of genomic resources- outside expressed sequence tags (ESTs) (Kantardjieff et al. 2009) which rendered engineering high producing cell lines more difficult. With the latest advances in sequencing and the publication of the CHO-K1 genome (see www.chogenome.org) this limitation is now being removed. At the time of writing this review a draft genome sequence and annotation of CHO-K1 sequence has been released publicly by Xu et al. (2011) and other sequencing efforts are also being undertaken internationally. Although the genome of CHO cell lines in other laboratories may diverge due to chromosomal rearrangements this resource remains a very important source of information and essential to allow the full utilisation and understanding of microRNA regulated gene expression in CHO cells. RNA sequencing in the future will further facilitate this field of research.

The production of complex biotherapeutics in a specific host also demands a good understanding of the dynamics that control the biology of the cell. As subtle environmental fluctuations are known to impact on the cell networks, it is crucial to develop methods to monitor the molecular changes that occur under specific conditions. The era of "omics" has therefore shed light on regulation at the gene, protein or metabolite levels (Clarke et al. 2011; Li et al. 2010, Melville et al. 2011; Nissom et al. 2006; Yee et al. 2008) but these studies only covered 10–15 % of the transcriptome and therefore a large amount of information remains missing or not investigated. In parallel to the various transcriptomic and proteomic studies that have been on-going in the field of CHO cell recombinant protein expression, the widespread impact of microRNAs has now been recognised and highlights a need for accurate, high-throughput techniques to quantify the abundance of each type of microRNA. As suggested previously in this chapter, deep sequencing refers to a ground-breaking method that generates millions of short RNA reads and allows the identification of microRNAs. However, it can be difficult to discriminate true microRNAs from fragments of other transcripts and short RNAs, especially when they are under-represented in the cell. The validation of a novel microRNA follows a set of rigorous rules such as the threshold for reads (> 10), sequences that flank the mature microRNA and lack of overlap with other annotated transcripts (Kozomara and Griffiths-Jones 2010).

As yet, only a handful of research groups have concentrated their efforts on identifying the miRNA population in CHO sub-types and across different culture conditions. For example, a study by Johnson et al (2010) isolated 350 miRs and classified them by running a BLAST alignment with known RNAs from the mIR-Base database. Their work at the time emphasised two important points: first, most microRNAs found in hamster are highly homologous to their human or mouse counterpart and secondly that a large number of miRNAs are more prone to fluctuations in their levels as a consequence of small environmental changes. Cgr-let-7f appears to be the most abundant microRNA in CHO cells and its levels were comparable across the conditions investigated in the reported study. This is not altogether surprising as the let-7f family is largely conserved. A similar approach was adopted by Hackl et al (2011) who mapped 387 mature microRNAs to an artificial sequence consisting of all microRNAs hairpins. Finally, Lin et al. (2011) compared the levels of miRNA in IgG producing and parental cell lines and found that miR-221 and miR-222 were significantly down-regulated in the antibody producing DG44 cell lines. However, there was no specific correlation between the volumetric IgG productivity and the relative expression of miR-221 and miR-222.

A further difficulty in CHO based microRNA studies remains that several of the target prediction software such as TargetScan still restrict their search to the human and mouse 3′UTRs and systematic investigation of targets for CHO miRNA is often hampered by the lack of an experimental database for the Chinese hamster. However, other software such as TargetRank allow the user to enter unknown sequences. If the degree of conservation of a particular miRNA between hamster and other vertebrates is high, one would predict that there is some degree of overlap in their target specificity. However, because of the different parameters and statistical techniques used in the current target prediction software it is sometimes difficult to define the ranking and scoring of targets. As an example, if one looks in either the mouse or the human genome for the putative targets of let-7f: TargetScan (http://www.targetscan.org/) and TargetRank (http://genes.mit.edu/targetrank/) agree on the target genes ranked 1 and 3 but disagree on the target gene ranked 2. However, the outcome is similar whether the search is undertaken with the human or the mouse microRNA. On the other hand, the top three predicted targets of miR-17, a broadly conserved microRNA, do not match when the query is run using the same research tools. As discussed above, because of the disparate nature of the target prediction it is essential to validate the predicted miRNA:mRNA physical interaction experimentally.

Investigators of CHO microRNA biology can use a number of approaches to validate the influence of perturbation on cell phenotype as discussed above. If a given mRNA is a true target of a specific miRNA, then perturbation of the mRNA:miRNA ratio should lead to a change in the amount of protein encoded by the target mRNA. Therefore, routine techniques such as Western analysis can be used to validate the functional importance of a given miRNA/mRNA target pair. Alternatively, ELISA experiments could also be utilized to quantify differences in protein expression. The target gene levels can also be verified by real-time RT-PCR. If the mRNA level is reduced, this suggests that the mechanism of protein downregulation involves mRNA degradation by deadenylation and exonucleolytic attack. However, if the miRNA does not primarily regulate gene expression by degradation of its target it may be more difficult to detect changes in mRNA levels but protein levels should still be altered due to translation attenuation.

To assess the impact of miR manipulation on protein expression, a few miRNA target validation reporter gene systems based on luminescence (Firefly (Tanguay and Gallie 1996), Renilla (Kong et al. 2008) or Gaussia luciferase (Kim et al. 2008)) have been developed for the quantitative assessment of the degree of post-transcriptional repression. Commonly, the 3′UTR of the predicted target is cloned downstream of the luciferase marker and the construct is transfected in the host cells with varying amounts of microRNAs (which can be upregulated from plasmid DNA or reduced using sponge vectors). Finally, the quantity of light emitted upon substrate cleavage is measured for each sample.

As alluded to previously, another addition to the repertoire of RNA-based gene regulation is the use of artificial transcripts, termed "sponges" or "antagomirs" that titrate endogenous microRNAs away from their target and make it less likely to bind and repress it. The first miRNA sponges were isolated in plants but these decoys systems could have many applications in the field of recombinant protein production as a tool in quality control. For example, by regulating genes involved in stress response, (Ebert et al. 2007; Ebert and Sharp 2010b; Ebert and Sharp 2010a).

1.7 Conclusion and Perspectives

Unravelling the influence of microRNAs on recombinant protein production in CHO cells is only just beginning. As such, despite the classification of the CHO microRNA library and the finding of a few miR signatures, at this stage it remains open to question as to whether microRNA levels are strongly linked to the recombinant protein production machinery and if their manipulation can be harnessed to develop improved CHO host cell lines. The yield of recombinant proteins currently observed in cultivated CHO cell lines often outperform dedicated secretory cells that live in the body (Wurm 2004). These yields reflect rapid advances in the culturing environment, expression vectors, host cell line development and bioengineering over the past 1–2 decades. As researchers begin to understand the biology of microRNA and the control these exert over gene expression new RNA-based approaches are being explored to potentially engineer hosts with further improved phenotypes. However, the diverse number of targets that microRNAs influence and the fact that miR levels change across culture means that manipulation of microRNAs to reliably influence improved phenotypes (e.g. growth, recombinant protein synthesis) is likely to be challenging and will only become a reality once a better appreciation of the biology and targets of miRs in the CHO cell is more fully developed.

References

Arvey A, Larsson E, Sander C, Leslie CS, Marks DS (2011) Target mRNA abundance dilutes microRNA and siRNA activity. Mol Syst Biol 6:363

Barron N, Kumar N, Sanchez N, Doolan P, Clarke C, Meleady P, O'Sullivan F, Clynes M (2011) Engineering CHO cell growth and recombinant protein productivity by overexpression of miR-7. J Biotechnol 151:204–211

Bartel, DP (2009) MicroRNAs: target recognition and regulatory functions. Cell 136:215–233

Betel D, Wilson M, Gabow A, Marks DS, Sander C (2008) The microRNA.org resource: targets and expression. Nucleic Acids Res 36:D149–153

Borchert GM, Lanier W, Davidson BL (2006) RNA polymerase III transcribes human microRNAs. Nat Struct Mol Biol 13:1097–1101

Brennecke J, Hipfner DR, Stark A, Russell RB, Cohen SM (2003) bantam encodes a developmentally regulated microRNA that controls cell proliferation and regulates the proapoptotic gene hid in Drosophila. Cell 113:25–36

Bueno MJ, Perez De Castro I, Malumbres M (2008) Control of cell proliferation pathways by microRNAs. Cell Cycle 7:3143–3148

Cai X, Hagedorn CH, Cullen BR (2004) Human microRNAs are processed from capped, polyadenylated transcripts that can also function as mRNAs. Rna 10:1957–1966

Calin GA, Dumitru CD, Shimizu M, Bichi R, Zupo S, Noch E, Aldler H, Rattan S, Keating M, Rai K, Rassenti L, Kipps T, Negrini M, Bullrich F, Croce CM (2002) Frequent deletions and down-regulation of micro- RNA genes miR15 and miR16 at 13q14 in chronic lymphocytic leukemia. Proc Natl Acad Sci USA 99:15524–15529

Clarke C, Doolan P, Barron N, Meleady P, O'Sullivan F, Gammell P, Melville M, Leonard M, Clynes M (2011) Large scale microarray profiling and coexpression network analysis of CHO cells identifies transcriptional modules associated with growth and productivity. J Biotechnol 155:350–359

Cui Q, Yu Z, Pan Y, Purisima EO, Wang E (2007) MicroRNAs preferentially target the genes with high transcriptional regulation complexity. Biochem Biophys Res Commun, 352:733–738

Ebert MS, Sharp PA (2010a) Emerging roles for natural microRNA sponges. Curr Biol 20:R858–861

Ebert MS, Sharp PA (2010b) MicroRNA sponges: progress and possibilities. Rna 16:2043–2050

Ebert M S, Neilson JR, Sharp PA (2007) MicroRNA sponges: competitive inhibitors of small RNAs in mammalian cells. Nat Methods 4:721–726

Friedman RC, Farh KK, Burge CB, Bartel DP (2009) Most mammalian mRNAs are conserved targets of microRNAs. Genome Res 19:92–105

Fussenegger M, Bailey JE (1998) Molecular regulation of cell-cycle progression and apoptosis in mammalian cells: implications for biotechnology. Biotechnol Prog 14:807–833

Gammell P, Barron N, Kumar N, Clynes M (2007) Initial identification of low temperature and culture stage induction of miRNA expression in suspension CHO-K1 cells. J Biotechnol 130:213–218

Godfried Sie CP, Kuchka M (2011) RNA editing adds flavor to complexity. Biochemistry (Mosc) 76:869–881

Gommans WM (2011) A-to-I editing of microRNAs: Regulating the regulators? Semin Cell Dev Biol.

Hackl M, Jakobi T, Blom J, Doppmeier D, Brinkrolf K, Szczepanowski R, Bernhart SH, Siederdissen CH, Bort JA, Wieser M, Kunert R, Jeffs S Hofacker I L, Goesmann A, Puhler A, Borth N, Grillari J (2011) Next-generation sequencing of the Chinese hamster ovary microRNA transcriptome: Identification, annotation and profiling of microRNAs as targets for cellular engineering. J Biotechnol 153:62–75

Han J, Lee Y, Yeom KH, Kim YK, Jin H, Kim VN. (2004) The Drosha-DGCR8 complex in primary microRNA processing. Genes Dev 18:3016–3027

Hinske LC, Galante PA, Kuo WP, Ohno-Machado L (2010) A potential role for intragenic miRNAs on their hosts' interactome. BMC Genomics 11:533

Hsu CW, Juan HF, Huang HC (2008) Characterization of microRNA-regulated protein-protein interaction network. Proteomics 8:1975–1979

Huang Y, Gu X (2011) A study of the evolution of human microRNAs by their apparent repression effectiveness on target genes. PLoS One 6:e25034

Hutvagner G, McLachlan J, Pasquinelli AE, Balint E, Tuschl Tm, Zamore PD (2001) A cellular function for the RNA-interference enzyme Dicer in the maturation of the let-7 small temporal RNA. Science 293:834–838

Hwang HW, Wentzel EA, Mendell JT (2009) Cell-cell contact globally activates microRNA biogenesis. Proc Natl Acad Sci USA 106:7016–7021

Ibanez-Ventoso C, Vora M, Driscoll M (2008) Sequence relationships among C. elegans, D. melanogaster and human microRNAs highlight the extensive conservation of microRNAs in biology. PLoS One 3:e2818

Johnson KC, Jacob NM, Nissom PM, Hackl M, Lee LH, Yap M, Hu WS (2010) Conserved MicroRNAs in Chinese hamster ovary cell lines. Biotechnol Bioeng.

Kantardjieff A, Nissom PM, Chuah SH, Yusufi F, Jacob NM, Mulukutla BC, Yap M, Hu WS (2009) Developing genomic platforms for Chinese hamster ovary cells. Biotechnol Adv 27:1028–1035

Kertesz M, Lovino N, Unnerstall U, Gaul U, Segal E (2007) The role of site accessibility in microRNA target recognition. Nat Genet 39:1278–1284
Kim HJ, Kim YH, Lee DS, Chung JK, Kim S (2008) In vivo imaging of functional targeting of miR-221 in papillary thyroid carcinoma. J Nucl Med 49:1686–1693
Kim JY, Kim YG, Lee GM (2011) CHO cells in biotechnology for production of recombinant proteins: current state and further potential. Appl Microbiol Biotechnol 93:917–930
Kim VN, Han J, Siomi MC (2009) Biogenesis of small RNAs in animals. Nat Rev Mol Cell Biol 10:126–139
Kim YK, Kim VN (2007) Processing of intronic microRNAs. Embo J 26:775–783
Kiryu H, Teral G, Imamura O, Yoneyama H, Suzuki K, Asai K (2011) A detailed investigation of accessibilities around target sites of siRNAs and miRNAs. Bioinformatics 27:1788–1797
Koh TC, Lee YY, Chang SQ, Nissom PM (2009) Identification and expression analysis of miRNAs during batch culture of HEK-293 cells. J Biotechnol 140:149–155
Kong YW, Cannell IG, De Moor CH, Hill K, Garside PG, Hamilton TL, Meijer HA, Dobbyn HC, Stoneley M, Spriggs KA, Willis AE, Bushell M (2008) The mechanism of micro-RNA-mediated translation repression is determined by the promoter of the target gene. Proc Natl Acad Sci USA 105:8866–8871
Kozomara A, Griffiths-Jones S (2010) miRBase: integrating microRNA annotation and deep-sequencing data. Nucleic Acids Res 39:D152–157
Krek A, Grun D, Poy MN, Wolf R, Rosenberg L, Epstein EJ, Macmenamin P, Da Piedade I, Gunsalus KC, Stoffel M, Rajewsky N (2005) Combinatorial microRNA target predictions. Nat Genet 37:495–500
Lagos-Quintana M, Rauhut R, Meyer J, Borkhardt A, Tuschl T (2003) New microRNAs from mouse and human. Rna 9:175–179
Lau PW, Macrae IJ (2009) The molecular machines that mediate microRNA maturation. J Cell Mol Med 13:54–60
Lee J, Li Z, Brower-Sinning R, John B (2007) Regulatory circuit of human microRNA biogenesis. PLoS Comput Biol 3:e67
Lee RC, Feinbaum RL, Ambros V (1993) The C. elegans heterochronic gene lin-4 encodes small RNAs with antisense complementarity to lin-14. Cell 75:843–854
Lee Y, Ahn C, Han J, Choi H, Kim J, Yim J, Lee J, Provost P, Radmark O, Kim S, Kim VN (2003) The nuclear RNase III Drosha initiates microRNA processing. Nature 425:415–419
Lewis BP, Burge CB, Bartel DP (2005) Conserved seed pairing, often flanked by adenosines, indicates that thousands of human genes are microRNA targets. Cell 120:15–20
Li L, Xu J, Yang D, Tan X, Wang H (2010) Computational approaches for microRNA studies: a review. Mamm Genome 21:1–12
Lin N, Davis A, Bahr S, Borgschulte T, Achtien K, Kayser K (2011) Profiling highly conserved microrna expression in recombinant IgG-producing and parental chinese hamster ovary cells. Biotechnol Prog.
Melville M, Doolan P, Mounts W, Barron N, Hann L, Leonard M, Clynes M, Charlebois T (2011) Development and characterization of a Chinese hamster ovary cell-specific oligonucleotide microarray. Biotechnol Lett 33:1773–1779
Min H, Yoon S (2010) Got target? Computational methods for microRNA target prediction and their extension. Exp Mol Med 42:233–244
Nam S, Kim B, Shin S, Lee S (2008) miRGator: an integrated system for functional annotation of microRNAs. Nucleic Acids Res 36:D159–164
Nissom PM, Sanny A, Kok YJ, Hiang YT, Chuah SH, Shing TK, Lee YY, Wong KT, Hu WS, Sim MY, Philp R (2006) Transcriptome and proteome profiling to understanding the biology of high productivity CHO cells. Mol Biotechnol 34:125–140
Park SY, Lee JH, Ha M, Nam JW, Kim VN (2009) miR-29 miRNAs activate p53 by targeting p85 alpha and CDC42. Nat Struct Mol Biol 16:23–29
Qiu C, Wang J, Yao P, Wang E, Cui Q (2010) microRNA evolution in a human transcription factor and microRNA regulatory network. BMC Syst Biol 4:90

Ruby JG, Jan CH, Bartel DP (2007) Intronic microRNA precursors that bypass Drosha processing. Nature 448:83–86
Saj A, Lai EC (2011) Control of microRNA biogenesis and transcription by cell signaling pathways. Curr Opin Genet Dev 21:504–510
Sandberg R, Neilson JR, Sarma A, Sharp PA, Burge CB (2008) Proliferating cells express mRNAs with shortened 3' untranslated regions and fewer microRNA target sites. Science 320:1643–1647
Stern-Ginossar N, Gur C, Biton M, Horwitz E, Elboim M, Stanietsky N, Mandelboim M, Mandelboim O (2008) Human microRNAs regulate stress-induced immune responses mediated by the receptor NKG2D. Nat Immunol 9:1065–1073
Suzuki HI, Miyazono K (2010) Dynamics of microRNA biogenesis: crosstalk between p53 network and microRNA processing pathway. J Mol Med (Berl) 88:1085–1094
Tanguay RL, Gallie DR (1996) Translational efficiency is regulated by the length of the 3' untranslated region. Mol Cell Biol 16:146–156
Trummer E, Fauland K, Seidinger S, Schriebl K, Lattenmayer C, Kunert R, Vorauer-Uhl K, Weik R, Borth N, Katinger H, Muller D (2006) Process parameter shifting: Part II. Biphasic cultivation-A tool for enhancing the volumetric productivity of batch processes using Epo-Fc expressing CHO cells. Biotechnol Bioeng 94:1045–1052
Westholm JO, Lai EC (2011) Mirtrons: microRNA biogenesis via splicing. Biochimie 93:1897–1904
Wijnhoven BP, Michael MZ, Watson DI (2007) MicroRNAs and cancer. Br J Surg 94:23–30
Winter J, Jung S, Keller S, Gregory RI, Diederichs S (2009) Many roads to maturity: microRNA biogenesis pathways and their regulation. Nat Cell Biol 11:228–234
Wurm FM (2004) Production of recombinant protein therapeutics in cultivated mammalian cells. Nat Biotechnol 22:1393–1398
Xu X, Nagarajan H, Lewis NE, Pan S, Cai Z, Liu X, Chen W, Xie M, Wang W, Hammond S, Andersen MR, Neff N, Passarelli B, Koh W, Fan HC, Wang J, Gui Y, Lee KH, Betenbaugh MJ, Quake SR, Famili I, Palsson BO, Wang J (2011) The genomic sequence of the Chinese hamster ovary (CHO)-K1 cell line. Nat Biotechnol 29:735–741
Yang F, Zhang L, Wang F, Wang Y, Huo XS, Yin YX, Wang YQ, Zhang L, Sun SH (2011) Modulation of the unfolded protein response is the core of microRNA-122-involved sensitivity to chemotherapy in hepatocellular carcinoma. Neoplasia 13:590–600
Yang W, Chendrimada TP, Wang Q, Higuchi M, Seeburg PH, Shiekhattar R, Nishikura K (2006) Modulation of microRNA processing and expression through RNA editing by ADAR deaminases. Nat Struct Mol Biol 13:13–21
Yee JC, De Leon Gatti M, Philp RJ, Yap M, Hu WS (2008) Genomic and proteomic exploration of CHO and hybridoma cells under sodium butyrate treatment. Biotechnol Bioeng 99:1186–1204
Yoon SK, Kim SH, Lee GM (2003a) Effect of low culture temperature on specific productivity and transcription level of anti-4–1BB antibody in recombinant Chinese hamster ovary cells. Biotechnol Prog 19:1383–1386
Yoon SK, Yand SJ, Lee GM (2003b) Effect of low culture temperature on specific productivity, transcription level, and heterogeneity of erythropoietin in Chinese hamster ovary cells. Biotechnol Bioeng 82:289–298

Chapter 2
Biogenesis of Mammalian miRNA

Stephen L. Clarke, McKale R. Davis and Ramanjulu Sunkar

Abstract MicroRNA (miRNA) are small noncoding molecules approximately 22 nucleotides in length that mediate translational repression of target messenger RNA (mRNA) transcripts. miRNA play important roles in cellular differentiation and proliferation, regulation of cellular metabolism, and are associated with enhanced tumorigenesis observed with many cancers. These small RNA molecules function by binding with imperfect complementarity to sequences located in the 3′ untranslated region (UTR) of target mRNA. MiRNA are expressed from both intergenic or intronic regions of the genome in an RNA polymerase II-dependent manner and their expression is subject to both feed-forward and negative feedback regulatory loops. Nascent transcripts known as primary miRNA (pri-mRNA) undergo nuclear processing to generate ∼ 60–80 nt precursor miRNA (pre-miRNA) containing imperfect stem-loop structures. Pre-miRNA are subsequently exported out of the nucleus whereupon they are further processed by the RNase III enzyme Dicer resulting in the formation of a ∼20–24 nt miRNA duplex. Thermodynamic stability of each strand determines which strand of the miRNA duplex will be retained within the RNA-induced silencing complex (RISC) to function as the mature miRNA. This mature miRNA then guides the RISC to sequences within the 3′UTR of target mRNAs to repress protein translation. Due to the roles that miRNA have in regulating cellular proliferation and metabolism, they could potentially interfere with recombinant protein expression in cell-based systems.

Keywords MicroRNA · Biogenesis · Transcription · Translational repression

---

S. L. Clarke (✉) · M. R. Davis
Department of Nutritional Sciences, Oklahoma State University, Stillwater, OK, USA
e-mail: stephen.clarke@okstate.edu

M. R. Davis
e-mail: mckale.davis@okstate.edu

R. Sunkar
Department of Biochemistry & Molecular Biology, Oklahoma State University, Stillwater, OK, USA
e-mail: ramanjulu.sunkar@okstate.edu

N. Barron (ed.), *MicroRNAs as Tools in Biopharmaceutical Production*,
DOI 10.1007/978-94-007-5128-6_2, © Springer Science+Business Media Dordrecht 2012

2.1 Introduction

MicroRNAs (miRNAs) are small non-coding regulatory RNA molecules (~22-nt) that post-transcriptionally regulate gene expression by mechanisms involving translational repression and/or decreasing mRNA stability. For the purposes of this review, the focus will be on regulation of mammalian miRNA biogenesis and mechanism of action. Because an individual miRNA species can regulate many targets, and because of the heterogeneity observed in mature miRNA sequences, miRNA serve an essential role in controlling the expression of transcripts and proteins involved in cellular development, differentiation, and metabolism. Furthermore, alterations in either miRNA expression or the expression/activity of components required for miRNA processing and target recognition may contribute to the pathogenesis of a variety of cancers and other chronic diseases. Thus, characterizing the factors that are responsible for controlling miRNA biogenesis is beneficial for enhancing our understanding of mechanisms allowing cells to respond to constantly changing environmental conditions in addition to providing therapeutic opportunities to prevent or treat chronic diseases.

2.2 Principles of MicroRNA Biogenesis

miRNA-dependent regulation of gene expression is thought to control the expression of as much as 60 % of the protein-coding genes expressed in mammals (Friedman et al. 2009). miRNA represent about ~1–2 % of the eukaryotic microtranscriptome and have diverse roles in modulating cellular processes including cellular proliferation, differentiation, and metabolism in addition to being involved disease pathogenesis (Liu et al. 2008; Williams 2008; Esau et al. 2006; Farazi et al. 2011).

miRNAs are transcribed primarily in an RNA polymerase II-dependent process from intergenic regions of the genome or from introns and exons of protein-coding genes (Fig. 2.1) (Kim and Nam 2006; Lee et al. 2004). The genomic organization of miRNA genes is the subject of an excellent review article and will only be briefly described here (Olena and Patton 2010). Approximately one-third of miRNA genes are located in the introns of protein-coding genes suggesting that the expression of miRNA and mRNA may be coordinately regulated (Westholm and Lai 2011; Rodriguez et al. 2004; Baskerville and Bartel 2005). In mammals ~36–46 % of known miRNA are clustered as polycistronic genes (Griffiths-Jones et al. 2008). The expression of miRNA from these clusters suggests that multiple miRNAs affecting multiple targets could be produced in response to environmental or developmental cues. As an additional level of complexity, targets of polycistronic miRNA are often regulated in such a way that multiple protein-protein interactions may be regulated by the induction of miRNA expression and subsequent translational control (Yuan et al. 2009). Expression of miRNA transcripts may also play an important role in autoregulatory loops. For example, one of the targets of hsa-miR-613 is the nuclear hormone

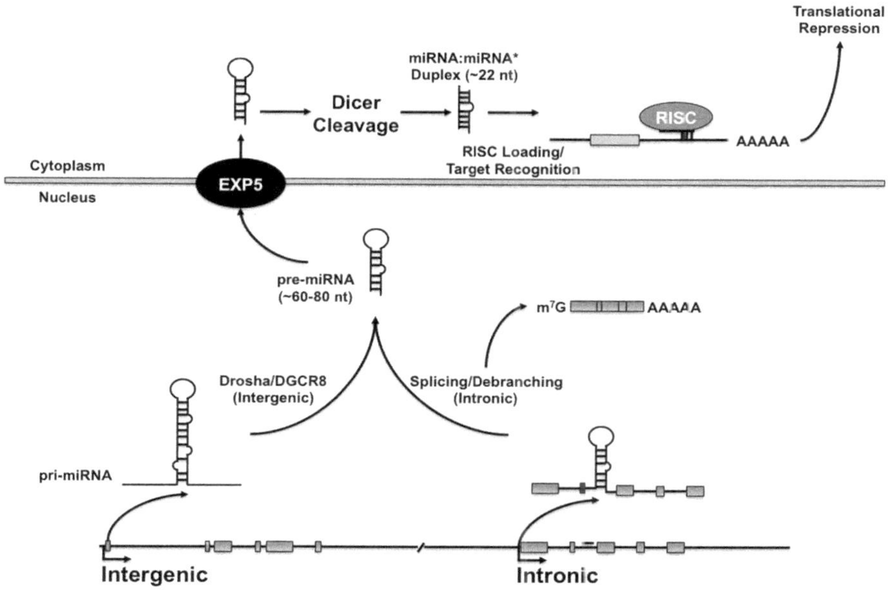

Fig. 2.1 Overview of mammalian miRNA biogenesis and function. The primary (pri-) miRNA transcripts that can adopt hairpin-like structures are transcribed from miRNA loci. Pri-miRNA transcripts from miRNA genes are processed to 60–80 nt pre-miRNA transcripts by a complex containing Drosha and DGCR8 in the nucleus. Alternatively, pre-miRNA may be derived from intronic regions of protein-coding genes in a Drosha/DGCR8 independent process requiring both the splicesome and a debranching enzyme known as the lariat debranching enzyme. Both the canonical Drosha-dependent processing and intronic processing pathways generate a pre-miRNA with a hairpin-like structure that is then exported out of the nucleus into the cytoplasm through Exportin 5 (EXP5). Along with Argonaute (Ago) proteins, Dicer processes the pre-miRNA transcript into a mature miRNA duplex. The strand in the duplex with the least thermodynamically stable 5′ end (guide strand) is retained by an Ago protein in mammals. The passenger strand (miRNA*) is generally released and degraded. Upon target recognition by the RNA-induced silencing complex (RISC) based on the seed region complementarity with the target mRNA, the target mRNA undergoes translational repression

receptor liver X receptor-alpha (LXRα) (Ou et al. 2011). The sterol response element binding protein (SREBP), a target of LXRα, binds an SREBP response element in the promoter of hsa-miR-613 and enhances its expression, thus down-regulating the expression of LXRα (Ou et al. 2011). Similarly, the expression of the clustered miRNA hsa-miR-2861 and hsa-miR-3960 is regulated by runt-related transcription factor 2 (Runx2), a key transcription factor that promotes osteoblast differentiation (Hu et al. 2011). Runx2 enhances the transcription of hsa-miR-2861 which results in the repression of histone-deacetylase 5 thereby inhibiting the repression of Runx2 to further promote osteoblast differentitation (Li et al. 2009). Runx2 also enhances the expression of hsa-miR-3960. A target of hsa-miR-3960 is homeobox a2 (Hoxa2), a protein that suppresses Runx2 expression (Hu et al. 2011). Thus, the

Runx2-mediated increase in miR-2861 and miR-3960 expression plays a key role in promoting osteoblast differentiation by suppressing inhibitors of Runx2 expression.

Similar to other protein-coding transcripts, these miRNA transcripts known as primary (pri-) miRNA undergo processing that promotes involving the addition of a 5′-m^{7}G cap structure and polyadenylation (Cai et al. 2004). These pri-miRNA transcripts are then processed to a ∼65-nt precursor (pre-) miRNA in the nucleus by the "Microprocessor" complex consisting of the RNase III-like enzyme Drosha and its co-factor DiGeorge Syndrome Critical Region 8 (DGCR8; Pasha in *D. melanogaster* and *C. elegans*) (Lee et al. 2003; Denli et al. 2004). Pre-miRNA containing hairpin-like structures are exported from the nucleus in a RanGTP-dependent process involving exportin 5 (Exp5) before undergoing additional cytosolic processing by Dicer to generate a mature miRNA (Bohnsack et al. 2004; Lund et al. 2004; Yi et al. 2003).

2.3 Nuclear Processing by Drosha

Drosha is a nuclear protein of 130–160 kDa containing two RNase III domains (RIIID) and a double-stranded RNA binding domain (dsRBD). Mammalian pri-miRNA transcripts contain dsRNA hairpins consisting of a ∼ 33 base pair (bp) stem, a terminal loop, and 5′ and 3′ flanking ssRNA. Although Drosha exhibits RNase III activity in the absence of its co-factor DGCR8, it requires DGCR8 in order to bind target RNA (Han et al. 2004). Instead, Drosha interacts with DGCR8 to form the core Microprocessor complex that is competent for RNA binding and cleavage. DGCR8 recognizes the ssRNA-dsRNA junction of target RNA through two dsRBDs and facilitates substrate cleavage ∼ 11 bp from the junction (Han et al. 2006; Zeng and Cullen 2005). The two RIIIDs of Drosha cleave the 5′ and 3′ strands of the stem resulting in the formation of a ∼ 65-nt (60–80-nt) pre-miRNA with a 2-nt 3′ overhang (Basyuk et al. 2003). The flanking sequence containing the 3′ overhang is an important factor for recognition by EXP5 and subsequent nuclear export for additional cytosolic processing (Bohnsack et al. 2004; Lund et al. 2004; Yi et al. 2003). After binding the pre-miRNA, the EXP5 co-factor GTP-bound nuclear Ran undergoes cytoplasmic hydrolysis releasing the pre-miRNA into the cytosol.

2.4 Cytosolic Processing by Dicer

Following generation of the pre-miRNA by the Microprocessor complex, the RNase III-like protein Dicer generates an RNA duplex containing the mature miRNA. Dicer, an ∼ 200 kDa protein containing DEAD-box RNA helicase domain, PAZ domain, two RIIIDs, and a dsRBD, cleaves the pre-miRNA near the loop to generate an ∼ 22-nt duplex that contains both the mature miRNA (guide strand) and a similarly sized RNA (passenger strand) derived from the opposite stem of the hairpin (Bernstein et al. 2001; Grishok et al. 2001; Hutvagner et al. 2001; Macrae et al. 2006).

Similar to Drosha in the nucleus, Dicer interacts with other proteins forming larger complexes involved in loading of the RNA induced silencing complex (RISC). The dsRNA binding protein TAR RNA binding protein (TRBP) and protein activator of PKR (PACT) are not required for pre-miRNA processing by Dicer, but contribute instead to the stability of the complex and formation of the RISC (Haase et al. 2005; Lee et al. 2006; Chendrimada et al. 2005; Melo et al. 2009; Paroo et al. 2009). The mature miRNA duplex is therefore determined both by recognition of the pri-miRNA by Drosha in the nucleus and cleavage by Dicer in the cytosol. The mature miRNA may be located on either side of the pre-miRNA hairpin duplex.

2.5 Targeting is Influenced by The 5′ End of The MicroRNA

The selection of cleavage sites by Drosha and Dicer may produce miRNA with multiple isoforms depending on the sequence of the 5′ and 3′ ends (Carthew and Sontheimer 2009). As explained below, differences in nucleotide composition can alter the "seed" sequence ultimately affecting which protein-coding mRNAs are targeted by the mature miRNA (Chiang et al. 2010). The ability of miRNA to repress translation is largely a function of the thermodynamic stability of base-pairing between the miRNA and complementary sequences generally found in the 3′UTR of target mRNA. In particular, nucleotides at positions 2–8 at the 5′ end of the miRNA comprise the so-called "seed" sequence for miRNA target recognition (Doench and Sharp 2004). Although both the 5′ and 3′ ends of miRNA can vary depending on the addition or deletion of 1–2 nt, alterations in the 5′ sequence can alter the seed sequence resulting in potential changes in miRNA target specificity. Additionally, thermodynamic stability of the 5′ and 3′ ends may affect which strand of the duplex is incorporated into the RISC (Krol et al. 2010). Differential selection of cleavage sites by Drosha and Dicer can therefore alter both the stability of the duplex and determine which strand is incorporated into the RISC as a mature miRNA.

2.6 RISC Assembly

Following cleavage by Dicer, the ~22-nt RNA duplex containing both the mature miRNA (guide strand) and the passenger strand (also referred to as the miRNA* strand) is incorporated into the RISC. In order to function in miRNA-mediated regulation of gene expression, the duplex is separated and the guide strand is retained in the RISC whereas the passenger strand is removed. RISC assembly occurs in two steps: first, ~22-nt RNA duplexes are incorporated into Argonaute (Ago) proteins and second, RNA duplexes are unwound or separated within Ago proteins (Kawamata and Tomari 2010). Argonaute proteins may be divided into three groups based on sequence similarity and the nature of their substrates: the AGO subfamily, the Piwi subfamily, and the *C. elegans*-specific WAGO subfamily (Czech and Hannon

2011). This review will focus on the role of AGO proteins on the processing and function of miRNA in mammals. Although the precise mechanism of assembly in mammals is not yet known, RNA duplex incorporation into AGO proteins and ATP-dependent RISC assembly appears to be a Dicer -independent process in contrast to the well-characterized Dicer-dependent RISC assembly involving Ago2 in flies (Kawamata and Tomari 2010; Schwarz et al. 2003).

Strand selection by the RISC is a regulated process dependent upon the thermodynamic stability of the nucleotides at the 5′ termini at each end of the duplex (Schwarz et al. 2003; Khvorova et al. 2003). This asymmetric strand selection due to the relative stability of the 5′ end is known as the "asymmetry rule". The strand with the least stable 5′ end (miRNA or guide strand) is retained within the Ago-associated complex, whereas the passenger strand is unwound from the duplex and degraded (Kawamata and Tomari 2010; Schwarz et al. 2003; Khvorova et al. 2003). Recent deep sequencing studies in *Drosophila* revealed that miRNA* species for a few miRNAs are abundantly expressed and dynamically regulated suggesting that miRNA* might in fact play important gene regulatory roles in a cell-type or cell-stage specific manner (Okamura et al. 2008). Indeed, miRNA* are capable of regulating the expression of luciferase reporter constructs containing the target sites for miRNA* suggesting that the passenger strands likely play an important role in mammalian gene regulatory networks (Yang et al. 2011).

All four AGO proteins expressed in humans (AGO1–4) can bind miRNA duplexes containing central mismatches (Yoda et al. 2010). In contrast to flies, there does not appear to be strict small RNA sorting system involved in RISC assembly in mammals. As a result of mismatches in base-pairing in the duplex, the two strands are unwound by a cleavage-independent (i.e., nuclease-independent) mechanism (Kawamata et al. 2009). Whereas the passenger strand appears to be susceptible to rapid nucleolytic degradation, the guide strand is retained in the RISC. Upon the formation of a mature or active RISC, the mature miRNA is then capable of exhibiting a regulatory function by altering protein translation or mRNA stability.

2.7 MicroRNA Repress Protein Translation

Due to the combinatorial nature of 5′ sequence heterogeneity in the seed region of the miRNA as a result of pri- and pre-miRNA processing, strand selection, and AGO association, the appropriate target recognition by mature miRNA is essential to its ability to participate in regulatory networks. Furthermore, miRNA relative expression levels and subcellular localization contribute to the complexity of miRNA-dependent regulation of gene expression. Although all four AGO proteins in humans evolved as ribonucleases, only AGO2 retains its catalytic activity (i.e., Slicer activity) required for the cleavage of target mRNA (Meister et al. 2004). Active RISC containing either AGO 1, 3, or 4 loaded with guide miRNA is able to recognize target sequence in the 3′UTR of target mRNA resulting in translational repression but could also decrease mRNA stability by promoting transcript degradation (Bartel 2009; Eulalio et al. 2008; Chekulaeva and Filipowicz 2009). mRNA targets subject to translational repression and degradation may be localized to cytoplasmic processing (P-) bodies that are associated with both mature RISC and target mRNA (Parker and Sheth 2007; Berezhna et al. 2011; Pothof and van Gent 2011). By contrast to widely characterized translational inhibition, miRNA can also increase or activate translation of target mRNA in a cell-cycle dependent manner (Vasudevan et al. 2007; Niepmann 2009; Roberts et al. 2011). For example, AU-rich elements in the 3′UTR of Tumor Necrosis Factor-α mRNA that are usually associated with promoting enhanced mRNA decay can recruit miRNA to either mediate translational up-regulation under serum-starved condition or repress translation during cellular proliferation (Vasudevan et al. 2007). These results suggest that miRNA can control both translational repression and activation of target mRNA in a cell-cycle specific manner thereby allowing cells to adapt to changes in metabolic or environmental conditions.

As mentioned above, miRNA interact with target mRNA by a mechanism involving imperfect base-pairing. Although there is little tolerance for mismatches in the seed region (nt 2–8 of the mature miRNA), targets containing mismatches in the seed region have nonetheless been described (Vella et al. 2004; Lindenblatt et al. 2009). Mismatches in the central portion of the miRNA repress endonucleolytic cleavage of target mRNA while simultaneously allowing the formation of a mature RISC by increasing the rate of strand dissociation (Yoda et al. 2010; Kawamata et al. 2009). Rather than a one miRNA: one mRNA model of repression, there are usually multiple miRNA and/or target sequences required for maximal repression (Doench and Sharp 2004; Hon and Zhang 2007). While most miRNA target sites located in the 3′UTR serve as the basis of a number of target prediction tools (Lhakhang and Chaudhry 2011; Huang et al. 2011; Hsu et al. 2011; Liu et al. 2010; John et al. 2004), target sites within the 5′UTR have also been described (Forman and Coller 2010; Grey et al. 2010; Moretti et al. 2010).

mRNA whose translation are cap-independent or mRNA lacking a 5′-m^{7}G cap structure are less robustly repressed than those mRNA that are cap-dependent (Pillai et al. 2005; Mathonnet et al. 2007). Additionally, internal ribosome entry site (IRES)-containing mRNA are poorly repressed by miRNA (Pillai et al. 2005; Humphreys et al. 2005). These results suggest that optimal repression may be dependent on disrupting the interaction between the cap binding eukaryotic initiation factor 4E (eIF4E) and the cap structure. Although the mechanisms are still not fully understood, alterations in protein-protein interactions are likely to result in an arrest of translational initiation. For example, GW182, a protein that is a component of the RISC, may interact with the poly(A) binding protein (PABP) thereby preventing PABP from binding to the scaffold protein eIF4G resulting in a block in translation initiation (Wakiyama et al. 2007; Wang et al. 2008; Fabian et al. 2009). Interestingly, both polyadenylated and deadenylated mRNA are subject to miRNA-mediated repression suggesting the existence of both poly(A)-dependent and independent mechanisms (Pillai et al. 2005; Beilharz et al. 2009). Some of these mRNA can also be found in P-body subcellular compartments providing a location wherein the repressed mRNA may be subsequently decapped and degraded (Parker and Sheth 2007; Cannell et al. 2008). Nonetheless, one of the molecular mechanisms through which miRNA modulate protein abundance is through the repression of translation by impairing translation initiation.

Existing evidence also suggests that miRNA function to repress translation at a point following translational initiation. Using sucrose gradient fractionation to examine polysomal distribution of mRNA, results indicate that both synthetic and endogenous miRNA target messages may be associated with polysomes (Petersen et al. 2006; Nottrott et al. 2006; Gu et al. 2009). Furthermore, synthetic mRNA containing a cap-independent IRES may be subject to repression by miRNA and provides evidence that the regulatory step occurs at some point after the initiation of translation (Petersen et al. 2006). In fact, numerous miRNA have been found associated with polysomal fractions indicating that mature miRNA control protein expression at a point beyond initiation and is consistent with a model wherein miRNA may also bind to and repress target mRNA at a post-initiation step (Kim et al. 2004; Nelson et al. 2004; Maroney et al. 2006; Kong et al. 2008). The precise mechanism controlling the miRNA-induced repression via regulating elongation or termination steps remains unknown.

2.8 Target Regulation by Altering Transcript Stability

Regulation of post-transcriptional gene expression by miRNA also likely occurs through the modulation mRNA stability or turnover. Following recognition of the target mRNA, miRNA promote the deadenylation and decapping of the message resulting in its destabilization (Fabian et al. 2010). Deadenylation may be, at least in part, dependent on the interaction between PABP and GW182 of the RISC thereby impairing the interaction between PABP and eIF4G (Fabian et al. 2009). It does not appear that deadenylation alone results in enhanced decay, but instead suggests that deadenylation may be requirement for decapping and degradation by either a $3'$–$5'$ or $5'$–$3'$ exoribonuclease. Intriguingly, recent work using rabbit reticulocyte lysate suggests that the poly(A) tail, but not the $5'$-m^{7}G cap, of target mRNA is required for miRNA-mediated effects (Ricci et al. 2011). In fact, in these studies using rabbit reticulocyte lysates, translational repression was independent of either deadenylation or decreased mRNA stability. Despite examples of deadenylation and decay occurring to varying degrees, the extent to which these processes play an important role in mammalian cells remains unclear and may indicate that the primary means of regulation, at least in mammals, is through a mechanism involved in the regulation of translation initiation (Fabian et al. 2009).

2.9 Summary

Similar to protein-coding genes, miRNA gene expression is usually regulated through the recruitment of RNA polymerase II to produce a primary miRNA transcript. This nascent transcript undergoes nuclear processing to generate 60–80 nt pre-miRNA transcripts that are then exported out of the nucleus and further processed by the

RNase III enzyme Dicer to generate a miRNA duplex. The strand of the duplex with the least thermodynamically stable 5′ terminus is subsequently incorporated into the RISC and directed towards recognition of the 3′UTR of target mRNA. In mammals, miRNA regulate protein expression primarily by repressing protein translation. Thus, alterations in miRNA expression allow cells to fine-tune protein expression to more appropriately control cellular fate and metabolism. Although there is an enhanced appreciation for the roles of miRNA in regulating cellular fate and metabolism, much remains to be known about the mechanisms controlling aspects of miRNA expression, miRNA turnover, and stoichiometry of miRNA: target ratios required for target repression.

References

Bartel DP (2009) MicroRNAs: target recognition and regulatory functions Cell 136(2):215–233. doi:10.1016/j.cell.2009.01.002

Baskerville S, Bartel DP (2005) Microarray profiling of microRNAs reveals frequent coexpression with neighboring miRNAs and host genes. Rna 11(3):241–247. doi:10.1261/rna.7240905

Basyuk E, Suavet F, Doglio A, Bordonne R, Bertrand E (2003) Human let-7 stem-loop precursors harbor features of RNase III cleavage products. Nucleic Acids Res 31(22):6593–6597

Beilharz TH, Humphreys DT, Clancy JL, Thermann R, Martin DI, Hentze MW, Preiss T (2009) microRNA-mediated messenger RNA deadenylation contributes to translational repression in mammalian cells. PLoS One 4(8):e6783. doi:10.1371/journal.pone.0006783

Berezhna SY, Supekova L, Sever MJ, Schultz PG, Deniz AA (2011) Dual regulation of hepatitis C viral RNA by cellular RNAi requires partitioning of Ago2 to lipid droplets and P-bodies. Rna 17(10):1831–1845. doi:10.1261/rna.2523911

Bernstein E, Caudy AA, Hammond SM, Hannon GJ (2001) Role for a bidentate ribonuclease in the initiation step of RNA interference. Nature 409(6818):363–365. doi:10.1038/35053110

Bohnsack MT, Czaplinski K, Gorlich D (2004) Exportin 5 is a RanGTP-dependent dsRNA-binding protein that mediates nuclear export of pre-miRNAs. Rna 10(2):185–191

Cai X, Hagedorn CH, Cullen BR (2004) Human microRNAs are processed from capped, polyadenylated transcripts that can also function as mRNAs. Rna 10(12):1957–1966. doi:10.1261/rna.7135204

Cannell IG, Kong YW, Bushell M (2008) How do microRNAs regulate gene expression? Biochem Soc Trans 36(Pt 6):1224–1231. doi:10.1042/BST0361224

Carthew RW, Sontheimer EJ (2009) Origins and Mechanisms of miRNAs and siRNAs. Cell 136(4):642–655. doi:10.1016/j.cell.2009.01.035

Chekulaeva M, Filipowicz W (2009) Mechanisms of miRNA-mediated post-transcriptional regulation in animal cells. Curr Opin Cell Biol 21(3):452–460. doi:10.1016/j.ceb.2009.04.009

Chendrimada TP, Gregory RI, Kumaraswamy E, Norman J, Cooch N, Nishikura K, Shiekhattar R (2005) TRBP recruits the Dicer complex to Ago2 for microRNA processing and gene silencing. Nature 436(7051):740–744. doi:10.1038/nature03868

Chiang HR, Schoenfeld LW, Ruby JG, Auyeung VC, Spies N, Baek D, Johnston WK, Russ C, Luo S, Babiarz JE, Blelloch R, Schroth GP, Nusbaum C, Bartel DP (2010) Mammalian microRNAs: experimental evaluation of novel and previously annotated genes. Genes Dev 24(10):992–1009. doi:10.1101/gad.1884710

Czech B, Hannon GJ (2011) Small RNA sorting: matchmaking for Argonautes. Nat Rev Genet 12(1):19–31. doi:10.1038/nrg2916

Denli AM, Tops BB, Plasterk RH, Ketting RF, Hannon GJ (2004) Processing of primary microRNAs by the Microprocessor complex. Nature 432(7014):231–235. doi:10.1038/nature03049

Doench JG, Sharp PA (2004) Specificity of microRNA target selection in translational repression. Genes Dev 18(5):504–511. doi:10.1101/gad.1184404

Esau C, Davis S, Murray SF, Yu XX, Pandey SK, Pear M, Watts L, Booten SL, Graham M, McKay R, Subramaniam A, Propp S, Lollo BA, Freier S, Bennett CF, Bhanot S, Monia BP (2006) miR-122 regulation of lipid metabolism revealed by in vivo antisense targeting. Cell Metab 3(2):87–98. doi:10.1016/j.cmet.2006.01.005

Eulalio A, Huntzinger E, Izaurralde E (2008) Getting to the root of miRNA-mediated gene silencing. Cell 132(1):9–14. doi:10.1016/j.cell.2007.12.024

Fabian MR, Mathonnet G, Sundermeier T, Mathys H, Zipprich JT, Svitkin YV, Rivas F, Jinek M, Wohlschlegel J, Doudna JA, Chen CY, Shyu AB, Yates JR, 3rd, Hannon GJ, Filipowicz W, Duchaine TF, Sonenberg N (2009) Mammalian miRNA RISC recruits CAF1 and PABP to affect PABP-dependent deadenylation. Mol Cell 35(6):868–880. doi:10.1016/j.molcel.2009.08.004

Fabian MR, Sonenberg N, Filipowicz W (2010) Regulation of mRNA translation and stability by microRNAs. Annu Rev Biochem 79:351–379. doi:10.1146/annurev-biochem-060308-103103

Farazi TA, Spitzer JI, Morozov P, Tuschl T (2011) miRNAs in human cancer. J Pathol 223(2):102–115. doi:10.1002/path.2806

Forman JJ, Coller HA (2010) The code within the code: microRNAs target coding regions. Cell Cycle 9(8):1533–1541

Friedman RC, Farh KK, Burge CB, Bartel DP (2009) Most mammalian mRNAs are conserved targets of microRNAs. Genome Res 19(1):92–105. doi:10.1101/gr.082701.108

Grey F, Tirabassi R, Meyers H, Wu G, McWeeney S, Hook L, Nelson JA (2010) A viral microRNA down-regulates multiple cell cycle genes through mRNA 5'UTRs. PLoS Pathog 6(6):e1000967. doi:10.1371/journal.ppat.1000967

Griffiths-Jones S, Saini HK, van Dongen S, Enright AJ (2008) miRBase: tools for microRNA genomics. Nucleic Acids Res 36(Database issue):D154–158. doi:10.1093/nar/gkm952

Grishok A, Pasquinelli AE, Conte D, Li N, Parrish S, Ha I, Baillie DL, Fire A, Ruvkun G, Mello CC (2001) Genes and mechanisms related to RNA interference regulate expression of the small temporal RNAs that control C. elegans developmental timing. Cell 106(1):23–34

Gu S, Jin L, Zhang F, Sarnow P, Kay MA (2009) Biological basis for restriction of microRNA targets to the 3' untranslated region in mammalian mRNAs. Nat Struct Mol Biol 16(2):144–150. doi:10.1038/nsmb.1552

Haase AD, Jaskiewicz L, Zhang H, Laine S, Sack R, Gatignol A, Filipowicz W (2005) TRBP, a regulator of cellular PKR and HIV-1 virus expression, interacts with Dicer and functions in RNA silencing. EMBO Rep 6(10):961–967. doi:10.1038/sj.embor.7400509

Han J, Lee Y, Yeom KH, Kim YK, Jin H, Kim VN (2004) The Drosha-DGCR8 complex in primary microRNA processing. Genes Dev 18(24):3016–3027. doi:10.1101/gad.1262504

Han J, Lee Y, Yeom KH, Nam JW, Heo I, Rhee JK, Sohn SY, Cho Y, Zhang BT, Kim VN (2006) Molecular basis for the recognition of primary microRNAs by the Drosha-DGCR8 complex. Cell 125(5):887–901. doi:10.1016/j.cell.2006.03.043

Hon LS, Zhang Z (2007) The roles of binding site arrangement and combinatorial targeting in microRNA repression of gene expression. Genome Biol 8(8):R166. doi:10.1186/gb-2007-8-8-r166

Hsu JB, Chiu CM, Hsu SD, Huang WY, Chien CH, Lee TY, Huang HD (2011) miRTar: an integrated system for identifying miRNA-target interactions in human. BMC Bioinformatics 12:300. doi:10.1186/1471-2105-12-300

Hu R, Liu W, Li H, Yang L, Chen C, Xia ZY, Guo LJ, Xie H, Zhou HD, Wu XP, Luo XH (2011) A Runx2/miR-3960/miR-2861 regulatory feedback loop during mouse osteoblast differentiation. J Biol Chem 286(14):12328–12339. doi:10.1074/jbc.M110.176099

Huang J, Townsend C, Dou D, Liu H, Tan M (2011) OMIT: A Domain-Specific Knowledge Base for MicroRNA Target Prediction. Pharm Res 28(12):310–314. doi:10.1007/s11095-011-0573-8

Humphreys DT, Westman BJ, Martin DI, Preiss T (2005) MicroRNAs control translation initiation by inhibiting eukaryotic initiation factor 4E/cap and poly(A) tail function. Proc Natl Acad Sci U S A 102(47):16961–16966. doi:10.1073/pnas.0506482102

Hutvagner G, McLachlan J, Pasquinelli AE, Balint E, Tuschl T, Zamore PD (2001) A cellular function for the RNA-interference enzyme Dicer in the maturation of the let-7 small temporal RNA. Science 293(5531):834–838. doi:10.1126/science.1062961

John B, Enright AJ, Aravin A, Tuschl T, Sander C, Marks DS (2004) Human MicroRNA targets. PLoS Biol 2(11):e363. doi:10.1371/journal.pbio.0020363

Kawamata T, Seitz H, Tomari Y (2009) Structural determinants of miRNAs for RISC loading and slicer-independent unwinding. Nat Struct Mol Biol 16(9):953–960. doi:10.1038/nsmb.1630

Kawamata T, Tomari Y (2010) Making RISC. Trends Biochem Sci 35(7):368–376. doi:10.1016/j.tibs.2010.03.009

Khvorova A, Reynolds A, Jayasena SD (2003) Functional siRNAs and miRNAs exhibit strand bias. Cell 115(2):209–216

Kim J, Krichevsky A, Grad Y, Hayes GD, Kosik KS, Church GM, Ruvkun G (2004) Identification of many microRNAs that copurify with polyribosomes in mammalian neurons. Proc Natl Acad Sci U S A 101(1):360–365. doi:10.1073/pnas.2333854100

Kim VN, Nam JW (2006) Genomics of microRNA. Trends Genet 22(3):165–173. doi:10.1016/j.tig.2006.01.003

Kong YW, Cannell IG, de Moor CH, Hill K, Garside PG, Hamilton TL, Meijer HA, Dobbyn HC, Stoneley M, Spriggs KA, Willis AE, Bushell M (2008) The mechanism of micro-RNA-mediated translation repression is determined by the promoter of the target gene. Proc Natl Acad Sci U S A 105(26):8866–8871. doi:10.1073/pnas.0800650105

Krol J, Loedige I, Filipowicz W (2010) The widespread regulation of microRNA biogenesis, function and decay. Nat Rev Genet 11(9):597–610. doi:10.1038/nrg2843

Lee Y, Ahn C, Han J, Choi H, Kim J, Yim J, Lee J, Provost P, Radmark O, Kim S, Kim VN (2003) The nuclear RNase III Drosha initiates microRNA processing. Nature 425(6956):415–419. doi:10.1038/nature01957

Lee Y, Hur I, Park SY, Kim YK, Suh MR, Kim VN (2006) The role of PACT in the RNA silencing pathway. Embo J 25(3):522–532. doi:10.1038/sj.emboj.7600942

Lee Y, Kim M, Han J, Yeom KH, Lee S, Baek SH, Kim VN (2004) MicroRNA genes are transcribed by RNA polymerase II. Embo J 23(20):4051–4060. doi:10.1038/sj.emboj.7600385

Lhakhang TW, Chaudhry MA (2011) Current approaches to micro-RNA analysis and target gene prediction. J Appl Genet 53(2):149–158. doi:10.1007/s13353-011-0060-2

Li H, Xie H, Liu W, Hu R, Huang B, Tan YF, Xu K, Sheng ZF, Zhou HD, Wu XP, Luo XH (2009) A novel microRNA targeting HDAC5 regulates osteoblast differentiation in mice and contributes to primary osteoporosis in humans. J Clin Invest 119(12):3666–3677. doi:10.1172/JCI39832

Lindenblatt C, Schulze-Osthoff K, Totzke G (2009) IkappaBzeta expression is regulated by miR-124a. Cell Cycle 8(13):2019–2023

Liu H, Yue D, Chen Y, Gao SJ, Huang Y (2010) Improving performance of mammalian microRNA target prediction. BMC Bioinformatics 11:476. doi:10.1186/1471-2105-11-476

Liu X, Fortin K, Mourelatos Z (2008) MicroRNAs: biogenesis and molecular functions. Brain Pathol 18(1):113–121. doi:10.1111/j.1750-3639.2007.00121.x

Lund E, Guttinger S, Calado A, Dahlberg JE, Kutay U (2004) Nuclear export of microRNA precursors. Science 303 (5654):95–98. doi:10.1126/science.1090599

Macrae IJ, Zhou K, Li F, Repic A, Brooks AN, Cande WZ, Adams PD, Doudna JA (2006) Structural basis for double-stranded RNA processing by Dicer. Science 311(5758):195–198. doi:10.1126/science.1121638

Maroney PA, Yu Y, Fisher J, Nilsen TW (2006) Evidence that microRNAs are associated with translating messenger RNAs in human cells. Nat Struct Mol Biol 13(12):1102–1107. doi:10.1038/nsmb1174

Mathonnet G, Fabian MR, Svitkin YV, Parsyan A, Huck L, Murata T, Biffo S, Merrick WC, Darzynkiewicz E, Pillai RS, Filipowicz W, Duchaine TF, Sonenberg N (2007) MicroRNA inhibition of translation initiation in vitro by targeting the cap-binding complex eIF4 F. Science 317(5845):1764–1767. doi:10.1126/science.1146067

Meister G, Landthaler M, Patkaniowska A, Dorsett Y, Teng G, Tuschl T (2004) Human Argonaute2 mediates RNA cleavage targeted by miRNAs and siRNAs. Mol Cell 15(2):185–197. doi:10.1016/j.molcel.2004.07.007

Melo SA, Ropero S, Moutinho C, Aaltonen LA, Yamamoto H, Calin GA, Rossi S, Fernandez AF, Carneiro F, Oliveira C, Ferreira B, Liu CG, Villanueva A, Capella G, Schwartz S, Jr., Shiekhattar R, Esteller M (2009) A TARBP2 mutation in human cancer impairs microRNA processing and DICER1 function. Nat Genet 41(3):365–370. doi:10.1038/ng.317

Moretti F, Thermann R, Hentze MW (2010) Mechanism of translational regulation by miR-2 from sites in the 5' untranslated region or the open reading frame. Rna 16(12):2493–2502. doi:10.1261/rna.2384610

Nelson PT, Hatzigeorgiou AG, Mourelatos Z (2004) miRNP:mRNA association in polyribosomes in a human neuronal cell line. Rna 10(3):387–394

Niepmann M (2009) Activation of hepatitis C virus translation by a liver-specific microRNA. Cell Cycle 8(10):1473–1477

Nottrott S, Simard MJ, Richter JD (2006) Human let-7a miRNA blocks protein production on actively translating polyribosomes. Nat Struct Mol Biol 13(12):1108–1114. doi:10.1038/nsmb1173

Okamura K, Phillips MD, Tyler DM, Duan H, Chou YT, Lai EC (2008) The regulatory activity of microRNA* species has substantial influence on microRNA and 3' UTR evolution. Nat Struct Mol Biol 15(4):354–363. doi:10.1038/nsmb.1409

Olena AF, Patton JG (2010) Genomic organization of microRNAs. J Cell Physiol 222(3):540–545. doi:10.1002/jcp.21993

Ou Z, Wada T, Gramignoli R, Li S, Strom SC, Huang M, Xie W (2011) MicroRNA hsa-miR-613 targets the human LXRalpha gene and mediates a feedback loop of LXRalpha autoregulation. Mol Endocrinol 25(4):584–596. doi:10.1210/me.2010-0360

Parker R, Sheth U (2007) P bodies and the control of mRNA translation and degradation. Mol Cell 25(5):635–646. doi:10.1016/j.molcel.2007.02.011

Paroo Z, Ye X, Chen S, Liu Q (2009) Phosphorylation of the human microRNA-generating complex mediates MAPK/Erk signaling. Cell 139(1):112–122. doi:10.1016/j.cell.2009.06.044

Petersen CP, Bordeleau ME, Pelletier J, Sharp PA (2006) Short RNAs repress translation after initiation in mammalian cells. Mol Cell 21(4):533–542. doi:10.1016/j.molcel.2006.01.031

Pillai RS, Bhattacharyya SN, Artus CG, Zoller T, Cougot N, Basyuk E, Bertrand E, Filipowicz W (2005) Inhibition of translational initiation by Let-7 MicroRNA in human cells. Science 309(5740):1573–1576. doi:10.1126/science.1115079

Pothof J, van Gent DC (2011) Spatiotemporal Aspects of MicroRNA-Mediated Gene Regulation. Adv Exp Med Biol 722:75–85. doi:10.1007/978-1-4614-0332-6_5

Ricci EP, Limousin T, Soto-Rifo R, Allison R, Poyry T, Decimo D, Jackson RJ, Ohlmann T (2011) Activation of a microRNA response in trans reveals a new role for poly(A) in translational repression. Nucleic Acids Res 39(12):5215–5231. doi:10.1093/nar/gkr086

Roberts AP, Lewis AP, Jopling CL (2011) miR-122 activates hepatitis C virus translation by a specialized mechanism requiring particular RNA components. Nucleic Acids Res 39(17):7716–7729. doi:10.1093/nar/gkr426

Rodriguez A, Griffiths-Jones S, Ashurst JL, Bradley A (2004) Identification of mammalian microRNA host genes and transcription units. Genome Res 14(10A):1902–1910. doi:10.1101/gr.2722704

Schwarz DS, Hutvagner G, Du T, Xu Z, Aronin N, Zamore PD (2003) Asymmetry in the assembly of the RNAi enzyme complex. Cell 115(2):199–208

Vasudevan S, Tong Y, Steitz JA (2007) Switching from repression to activation: microRNAs can up-regulate translation. Science 318(5858):1931–1934. doi:10.1126/science.1149460

Vella MC, Choi EY, Lin SY, Reinert K, Slack FJ (2004) The C. elegans microRNA let-7 binds to imperfect let-7 complementary sites from the lin-41 3'UTR. Genes Dev 18(2):132–137. doi:10.1101/gad.1165404

Wakiyama M, Takimoto K, Ohara O, Yokoyama S (2007) Let-7 microRNA-mediated mRNA deadenylation and translational repression in a mammalian cell-free system. Genes Dev 21(15):1857–1862. doi:10.1101/gad.1566707

Wang B, Yanez A, Novina CD (2008) MicroRNA-repressed mRNAs contain 40 S but not 60 S components. Proc Natl Acad Sci U S A 105(14):5343–5348. doi:10.1073/pnas.0801102105

Westholm JO, Lai EC (2011) Mirtrons: microRNA biogenesis via splicing. Biochimie 93(11):1897–1904. doi:10.1016/j.biochi.2011.06.017

Williams AE (2008) Functional aspects of animal microRNAs. Cell Mol Life Sci 65(4):545–562. doi:10.1007/s00018-007-7355-9

Yang JS, Phillips MD, Betel D, Mu P, Ventura A, Siepel AC, Chen KC, Lai EC (2011) Widespread regulatory activity of vertebrate microRNA* species. Rna 17(2):312–326. doi:10.1261/rna.2537911

Yi R, Qin Y, Macara IG, Cullen BR (2003) Exportin-5 mediates the nuclear export of pre-microRNAs and short hairpin RNAs. Genes Dev 17(24):3011–3016. doi:10.1101/gad.1158803

Yoda M, Kawamata T, Paroo Z, Ye X, Iwasaki S, Liu Q, Tomari Y (2010) ATP-dependent human RISC assembly pathways. Nat Struct Mol Biol 17(1):17–23. doi:10.1038/nsmb.1733

Yuan X, Liu C, Yang P, He S, Liao Q, Kang S, Zhao Y (2009) Clustered microRNAs' coordination in regulating protein-protein interaction network. BMC Syst Biol 3:65. doi:10.1186/1752-0509-3-65

Zeng Y, Cullen BR (2005) Efficient processing of primary microRNA hairpins by Drosha requires flanking nonstructured RNA sequences. J Biol Chem 280(30):27595–27603. doi:10.1074/jbc.M504714200

Chapter 3
Target Prediction Algorithms and Bioinformatics Resources for miRNA Studies

Colin Clarke, Niall Barron, Mark Gallagher, Michael Henry, Paula Meleady and Martin Clynes

Abstract The recent publication of the Chinese hamster ovary (CHO) genome has heralded the beginning of an exciting new era of research in this industrially important cell line. Advances in our understanding of CHO at the molecular level have the potential to facilitate the development of modified cell lines and biomarkers to increase the efficiency of recombinant protein production processes. In recent years there has been growing interest in the function of small non-coding RNA molecules, known as microRNAs (miRNAs), as targets to enable multigene CHO cell engineering. To date, miRNAs have been shown to be dysregulated in a number of processes including cell growth and apoptosis.

Bioinformatics has proven to be an essential supporting technology for miRNA based studies. In this chapter, we review a new class of miRNA specific *in-silico* tool developed to predict which mRNAs a particular miRNA targets in order to determine the impact of a miRNA on biological function. A range of popular miRNA target prediction algorithms are presented, their underlying principles described and performance assessed. In addition, publically available repositories of miRNA sequence, expression profiling and target data are highlighted. Finally, examples of the utilisation of these tools to study CHO cells are presented.

Keywords MicroRNA · Bioinformatics · Target prediction algorithm · MicroRNA data repository · CHO · Bioprocess

3.1 Introduction

The study of miRNA-mediated regulation of CHO cells in industrial culture has progressed rapidly despite the fact that until recently no genome sequence was available (Xu et al. 2011). In contrast to other "omics" platforms, the high degree of conservation of miRNAs across mammalian species has allowed the utilisation of

---

C. Clarke (✉) · N. Barron · M. Gallagher · M. Henry · P. Meleady · M. Clynes
National Institute for Cellular Biotechnology, Dublin City University, Glasnevin, Dublin 9, Ireland
e-mail: colin.clarke@dcu.ie

N. Barron (ed.), *MicroRNAs as Tools in Biopharmaceutical Production*,
DOI 10.1007/978-94-007-5128-6_3,

cross-species microarrays (Gammell et al. 2007), qPCR and next generation sequencing (NGS) platforms (Hackl et al. 2011; Hammond et al. 2011; Johnson et al. 2011) to study these small RNAs in CHO. To date, miRNAs have been implicated in the control of several industrially relevant phenotypes including growth (Bort et al. 2011; Gammell et al. 2007), apoptosis (Druz et al. 2011), and recombinant protein production (Barron et al. 2011; Hammond et al. 2011; Lin et al. 2010). As we enter the next phase of CHO miRNA research computational tools will become increasingly important in unravelling the effects of miRNA expression on the transcriptome, proteome and ultimately bioprocess phenotypes.

Bioinformatics plays a crucial role in the study of miRNA biology. Researchers utilise a variety of supporting computational methods in a range of areas from sequence analysis to detecting differential expression of miRNAs. While pre-existing tools were readily available for some of these tasks, a new class of *in-silico* methods to predict miRNA interactions with mRNA (therefore predicting miRNA function) was required. The development of these algorithms was driven by the breadth and complexity of animal miRNA dependant post-transcriptional control (~ 30 % of all protein coding genes are estimated to be targeted by one or more miRNA (Filipowicz et al. 2008)) along with the extensive wet-lab experimentation required to confirm a miRNA target.

The aim of this chapter is to describe miRNA target prediction algorithms; we begin by describing the currently accepted rules of miRNA target interaction (Sect. 3.2), followed by a detailed treatment of the most popular algorithms (Sect. 3.3). Publicly available data repositories containing precompiled lists of predicted targets, validated targets and sequence information are described (Sect. 3.4); the chapter closes with a discussion of algorithm performance (Sect. 3.5) and examples of their application in CHO studies (Sect. 3.6).

3.2 Principles of MicroRNA Target Prediction

The prediction of miRNA targets in plants is generally straightforward. Here miRNAs tend to bind their mRNA targets with near perfect complementarity inducing cleavage of the transcript (Rhoades et al. 2002), and while near-perfect matches between metazoan miRNAs and their respective targets can be found, it is unusual. Prediction of target interactions in animals is generally more challenging due to the intricacy of target recognition where sequences often contain gaps, mismatches and G:U base pairs in multiple positions (Bartel 2009). The complexity of animal miRNA regulation is reflected by estimates that a single miRNA targets an average of 100–200 mRNAs (Krek et al. 2005), a single mRNA transcript can be targeted by hundreds of miRNAs (Friedman et al. 2009) and multiple miRNAs can cooperatively repress a range of targets (Wu et al. 2010). Experimental methodologies to confirm direct miRNA regulation have advanced over recent years; however the number of confirmed interactions with mRNA remains relatively small. Computational target prediction remains a critical aspect of assessing global miRNA function and

Fig. 3.1 miRNA-target recognition site types. The four canonical site types are shown for myotubularin related protein 3 mRNA 3′UTR and human mir-16–1

prioritising candidates for wet-lab confirmation (Orom and Lund 2010). This section aims to give the reader an overview of the sequence-based rules that underpin the most widely used target prediction algorithms from comparative sequence analyses to thermodynamics.

3.2.1 MicroRNA Seed Region Base Pairing

Determining the degree of complementarity between bases 2–7 at the 5′ end of the mature miRNA sequence (termed the seed region) to the target transcript (some algorithms require perfect Watson-Crick base pairing to the seed region) is the most widely used feature in computational target prediction (Bartel 2009). The importance of this region was first highlighted by Lewis and co-workers who demonstrated a greater degree of seed region conservation across several species than would be expected by chance (Lewis et al. 2003). Furthermore, studies that perturbed specific miRNAs followed by global expression profiling of mRNA (Lim et al. 2005) and proteins (Baek et al. 2008; Selbach et al. 2008) observed enrichment of seed region target sites within the resulting differentially expressed gene/protein sets. It has also been shown that mutations within the miRNA seed region can lead to an elimination of miRNA function, enhance or reduce their biogenesis (Sun et al. 2009) and result in the onset of disease (Hughes et al. 2011; Mencia et al. 2009) The extent of gene expression repression is thought to be affected by variations in the number of contiguous bases and sequence composition of the gene target binding the seed (Fig. 3.1). It also should be noted that, in some instances, miRNA mediated repression can occur in the absence of a seed region match. These "seedless" highly complementarity miRNA-mRNA interactions have also been shown to mediate repression (Azzouzi et al. 2011; Lal et al. 2009). The type of seed match however, remains central to the scoring systems of many of the algorithms mentioned later in this chapter (Sect. 3.3).

The most frequently occurring type of seed regions are known as canonical target sites and are broadly defined by continuous base pairing spanning 7 or 8 nucleotides. These site types can be further characterised by a match at position 8 (*7mer-m8 site*), a 7 nucleotide match with binding to an adenosine at position 1 (*7mer-A1 site*), and matching either any base at position 8 or an adenosine matched at position 1 at the miRNA 5′ end (*8mer site*) (Bartel 2009). Studies have demonstrated that an adenosine at position 1 (complementary or non-complementary) increases down regulation of the target in comparison to other target site types (Baek et al. 2008; Nielsen et al. 2007). The presence of adenosines around the seed region seems to be conserved (Lewis et al. 2005) and is thought to be important for recognition by the RNA induced silencing complex (RISC).

Another class of seed region types can be categorised by an additional 4–5 base complementarity with the target mRNA towards the miRNA 3′ end (Bartel 2009). These types of target recognition have either a perfect match to the seed region (*3′-supplementary*) or correct for a mismatch in the seed region (*3′-compensatory*). Binding of nucleotides at positions 13–16 of the mature miRNA seems to provide optimum down regulation of targets (Grimson et al. 2007). The final seed type is termed "marginal" and is defined by only 6 nucleotide matches with the seed region (*6mer site*). Base pairing can begin at position 2 or position 3 of the miRNA. Those algorithms requiring perfect Watson Crick base pairing with the seed region in order to decrease the false positive rate would therefore ignore these marginal seeds (Bartel 2009).

3.2.2 mRNA Target Site Location and Abundance

The vast majority of metazoan miRNA binding sites reside within the 3′ UTR of the target transcript. The first miRNAs discovered, lin-4 and let-7 in *C. elegans*, both bind to sites within the 3′UTR of their respective targets (Lee et al. 1993; Reinhart et al. 2000). The presence of multiple target sites for one or more miRNAs within the 3′ UTR tends to increase miRNA induced repression. In addition, if these sites are within 13–35 bases of each other they can act in unison to further downregulate target translation (Brennecke et al. 2005; Saetrom et al. 2007). Target gene sites can be distributed toward both ends of the 3′UTR. The location of the target site along with surrounding sequence composition of the target site seems to affect the degree of miRNA regulation. For instance, characteristics of target sites including lying within 15 nt of the stop codon, away from the centre of long UTRs, near coexpressed miRNAs and close to AU rich regions, tends to increase effectiveness (Grimson et al. 2007). Kertesz et al. provided evidence of the importance of target site context, demonstrating the impact of target site accessibility and mRNA secondary structure on the efficiency of translation repression via miRNA. The importance of accessibility within the 3′UTR was further underlined by the observation that these functional target sites seem to be under evolutionary pressure (Kertesz et al. 2007).

A large proportion of target algorithms, therefore, continue to place emphasis on the 3′ end of the transcript during scoring. However these sites are not solely confined to the 3′UTR and recent evidence points to biologically active sites located within both the 5′UTR and coding sequence (CDS) (Moretti et al. 2010; Schnall-Levin et al. 2011). Sites within these regions seem to occur less frequently and are on the whole less effective than those within 3′UTR (Bartel 2009), although target binding sites within the 5′UTR of the transcript can in some cases disrupt translation as efficiently as those within the 3′UTR (Lytle et al. 2007) and in some instances simultaneous targeting of both the 5′ and 3′UTR by a single miRNA can also occur (Lee et al. 2009).

3.2.3 Cross Species Conservation

The number of common miRNA targets predicted by the early algorithms following analysis of the same species was low. To alleviate these concerns a further step was introduced to determine evolutionary conservation across multiple species which allowed targets that scored equally well in sequence alignment to be ranked according to conservation (Bartel 2009). The rationale behind the use of conservation analysis of transcripts and miRNA target regions is that miRNA-mRNA interactions that are present across a large evolutionary distance suggests biological functionality. The majority of target algorithms utilise a combination of seed region pairing and phylogenetic filtering to reduce false positive results. While conservation provides a strong indication for functional relevance, not all functional targets of miRNAs are conserved (Farh et al. 2005).

3.2.4 Thermodynamic Calculations

In recent years significant false positive rates associated with seed rule/conservation based approaches have driven the development of algorithms that assess the binding efficiency of a miRNA with its target. Following the observation that the free energy of hybridisation between the miRNA and mRNA target site (ΔG_{duplex}) was an important factor in determining biological function (Doench and Sharp, 2004), thermodynamic calculations were incorporated into several algorithms. ΔG_{duplex} can be calculated for each predicted miRNA-mRNA pair, the most commonly used program for this task is known as the Vienna package (http://www.tbi.univie.ac.at/RNA/) (Hofacker 2003). The lower the free energy the greater the likelihood of an interaction: a ΔG_{duplex} threshold can be applied to remove putative target interactions above a free energy level. miRNA target prediction schemes tend to vary this threshold according to species, e.g. *Drosophila*—14 kcal/mol (Enright et al. 2003), Human—17 kcal/mol

(John et al. 2004) and *C. elegans*—20 kcal/mol (Watanabe et al. 2006). Kertesz et al have extended the principle of thermodynamic feasibility through the calculation of the free energy of target accessibility (ΔG_{open}) governing the "opening" of mRNA structure (target site along with upstream and downstream flanking sequences) to facilitate the binding of miRNA to target (Kertesz et al. 2007).

3.3 miRNA Target Prediction Algorithms

3.3.1 miRanda

The miRanda algorithm was originally developed for target prediction in *D. melanogaster* (Enright et al. 2003) and has now been extended to include human, mouse, rat and *C. elegans* (Betel et al. 2008; John et al. 2004). miRanda initially assesses the complementarity of an entire miRNA to each mRNA 3′UTR via an extension of the classical Smith-Waterman alignment method. miRanda requires an almost perfect match within the seed region, although wobble is permitted and can be compensated by an extra match at the 3′end of the miRNA.

A sequence position dependent "weighted" alignment score (S) is calculated for each potential mRNA target. Matches in the seed region of the miRNA are assigned a higher weighting. The algorithm penalises G-U wobbles, gaps, insertions and deletions. Scoring also takes into account the proximity of the target site to the miRNA 5′ end. The best scoring alignments are filtered on thermodynamic feasibility by calculating the ΔG_{duplex} of the putative miRNA-mRNA interaction using the Vienna algorithm (Wuchty et al. 1999). miRanda retains only interactions when target sequences are above an evolutionary conservation threshold (e.g. PhastCons score > 0.57) and the target site reside at approximately the same sequence position (Betel et al. 2008; Enright et al. 2003).

3.3.2 PicTar

The PicTar (probabilistic identification of target sites) (Krek et al. 2005) algorithm differs from other target prediction strategies in that target sites for multiple miRNAs as well as single miRNAs can be identified. PicTar, inspired by the observation of coexpression between miRNAs and the presence of multiple target sites within 3′UTRs (Hobert, 2004), attempts to reduce false positives through the identification of common targets of coexpressed miRNAs. Target identification begins with a set of miRNAs along with multiple orthologous 3′UTR sequence alignments and followed by identification of perfect complementarity with the seed region or disruption of complementarity by a single base difference (bulge, mismatch or wobble) at most. These matches within regions of the 3′UTR termed "nuclei" are subjected to thermodynamic filtering of the resulting duplexes. A free energy threshold is applied, for instance a complimentary nuclei the free energy of binding must be within 33 % of the optimum free energy, while imperfect nuclei must be within 66 %. The remaining results are scored against a Hidden Markov Model (HMM) based ranking of the probability of being common targets of subsets of the miRNAs under consideration. Finally conservation analysis across eight vertebrate species is utilised to prioritise targets and eliminate false positives. The scores for each species are calculated individually and combined to give a final score for a gene. The usefulness of this algorithm is demonstrated by PicTar prediction and subsequent experimental confirmation of miR-375, miR-124 and let-7b coregulation of the myotrophin gene (Krek et al. 2005).

3.3.3 Diana-microT

The Diana-microT (Maragkakis et al. 2009a, b) approach utilises an extended miRNA seed region of up to nine nucleotides known as the "miRNA driver sequence". This driver sequence is shifted across the 3′UTR of the target gene to search for alignments with 7, 8 or 9 nt Watson-Crick base-pairing. DIANA-microT also aims to incorporate sites with weaker binding of 6 nts or G:U wobbles within target prediction. DIANA-microT 3.0 attempts to control false positives by filtering potential target sites without 7 nt Watson-Crick base pairing by applying a threshold to thermodynamic calculations provided by RNAhybrid (See below). Each individual miRNA binding site is scored with respect to the binding type (e.g. 6mer) and conservation. These scores are combined for each individual 3′UTR to produce a gene level "miTG score" (incorporating assessment against a set of synthetic miRNAs).

3.3.4 TargetScan/TargetScanS

TargetScan was the first method developed to predict miRNA targets in humans (Lewis et al. 2003). The underlying rationale of the TargetScan method is that, by returning only perfect seed matches and utilising only groups of orthologous 3′UTRs as input, the false positive rate can be controlled (although non-conserved sites will be missed). The algorithm begins by locating perfect complementarity to the miRNA seed regions. Once a seed match is located TargetScan extends the alignment along the 3′UTR until a mismatch is located (G:U wobbles are permitted). The optimum base pairing at the 3′ end of the miRNA is determined and RNAfold calculates the free energy to the miRNA-mRNA interaction. For each alignment a "Z-score" incorporating thermodynamic stability is calculated for each species to assign a rank to each potential miRNA-target interaction. Further refinement produced the simplified TargetScanS (Lewis et al. 2005) which is a simplification of TargetScan algorithm where the criteria for a target match is less stringent and only complementarity to a 6 nt seed match along with an adenosine located at position 1 is required. The algorithm also ranks the outputted prediction based on a "context + score" calculated from the target type, AU composition, position, site abundance and seed pairing stability (Garcia et al. 2011; Grimson et al. 2007).

3.3.5 PITA

The PITA algorithm extends the thermodynamic calculation to consider target accessibility with respect to transcript secondary structure (Kertesz et al. 2007). The initial stage of PITA predicts targets via seed region complementarity to the mRNA 3′UTR over at least seven bases for human, mouse, worm and fly followed by the application of a thermodynamic model to calculate $\Delta\Delta G$, a comparison of the free energy gained from binding of the miRNA to the potential target gene ΔG_{duplex} and the cost of un-pairing the target for miRNA accessibility ΔG_{open}. The user can also limit results to conserved miRNAs and mRNA if desired.

3.3.6 RNAhybrid

RNAhybrid attempts to locate the most likely target site with the miRNA seed through the identification of thermodynamically stable matches in the mRNA 3′UTR (Kruger and Rehmsmeier 2006; Rehmsmeier et al. 2004). RNAhybrid does not assess pre-computed alignments to the seed region but rather identifies any regions that have the ability to form duplexes with a given miRNA. The algorithm also applies a statistical adjustment to correct for the sequence length of the target and miRNA sequence and assigns a p-value. In addition, the significance of multiple and conserved sites can be computed using Poisson statistics.

3.3.7 RNA22

The RNA22 algorithm (Miranda et al. 2006) differs from competing methods in that target prediction does not use evolutionary filtering allowing the identification of species specific miRNA-mRNA interactions. In addition, RNA22 does not rely on locating the reverse complement of the miRNA seed region but attempts to find overrepresented statistical patterns derived from the mature miRNA within the transcript of interest potentially identifying unknown miRNA binding sites (RNA22 can assess the 3′UTR, 5′UTR and CDS). The initial phase of RNA22 identifies targets sites by identifying motifs within the mature miRNA. The reverse complement of statistically significant motifs are compared to the transcript to locate potential binding sites. As the patterns located in stage one of the algorithm contain redundancy, one or more of these patterns can map to a single transcript. If a pattern aligns at a particular location a vote is cast and if multiple patterns align to a location (e.g. votes ≥ 30) the area is identified as a "target island" for potential miRNA binding. To associate a mature miRNA (as opposed to mature miRNA pattern) those areas with low pattern aggregation are discarded and the remaining target islands are aligned to the miRNA sequences in a straightforward manner followed by thermodynamic assessment of binding via the Vienna package after insertion of a linker to form an *in-silico* hetroduplex.

3.3.8 mirWalk

mirWalk (Dweep et al. 2011) attempts to locate putative miRNA-mRNA interactions not only in 3′UTR but in any region of the mRNA. Once Watson-Crick base pairing across 7 nucleotides is located the miRNA seed is extended until a mismatch is found. For each miRNA the targets matched are categorised in terms of their position within the transcript (promoter region, 5′UTR, coding sequence, 3′UTR or mitochondrial genes) with a p-value reported for each match based on a random seed match. The mirWalk system allows user friendly overlap with several other prediction algorithms as well as a collection of validated targets.

3.3.9 Machine Learning

Several approaches have recently emerged that utilise features from known miRNA-mRNA interactions. Such methods aim to develop a statistical model to predict novel interactions from the characteristics of those that have already been discovered. One such approach was developed by Kim and co-workers (Kim et al. 2006). A support vector machine approach (SVM) algorithm was applied to predict targets for multiple species. The data utilised to train the algorithm on features selected from a combination of sequence composition and binding position along with thermodynamic calculations. miRSVR is an example of a similar approach where a model was constructed from 57 features derived using a two-class Natrive Bayes (NBmiRTar) (Yousef et al. 2007) (an artificial negative target dataset was generated for training). The model was integrated with the predicted targets of miRanda demonstrating how machine learning and target prediction strategies can be used together. Several other machine learning approaches have also been described using a variety of competing algorithms including random forest prediction (Jiang et al. 2007), and in some instances a combination of different machine learning algorithms (Yan et al. 2007).

3.3.10 Combining in-silico Prediction with Experimental Data

One recently developed approach, HOCTAR (Host Gene Oppositely correlated Targets) (Gennarino et al. 2009) utilises gene expression analysis to search for inversely correlated target genes. The creators of the algorithm compiled at list of predicted targets outputted by TargetScan, miRanda and PicTar. The intergenic miRNAs within each of the lists were compiled and a single list of miRNAs produced. A dataset consisting of $\sim$3500 microarray samples was also compiled to calculate correlation values. The probesets on the chip targeting a miRNA host gene were identified. For each host gene the Pearson correlation coefficient was calculated between the gene and all other genes on the array for each dataset. The 3 % of genes with the largest inverse correlation from each dataset were compiled into a list ranked on the number of times a gene appeared. The HOCTAR method suffers from the fact that miRNAs under investigation must result in mRNA transcript degradation and those targets undergoing translation repression would not be detected. A database of HOCTAR target predictions is available and can be searched by miRNA or gene name (Gennarino et al. 2011).

3.4 Publicly Available miRNA Data Repositories

3.4.1 miRbase

miRbase is a web-based, searchable database designed to provide access to miRNA sequences and annotation along with links to miRNA expression profiling data via NCBI GEO (Kozomara and Griffiths-Jones 2011). The database contains both hairpin precursor miRNA and mature miRNA sequences. At the time of writing the latest version, mirBase 18 contained 18,226 hairpin precursors and 21,643 mature miRNA sequences from 168 species.

3.4.2 microRNA.org

microRNA.org (Betel et al. 2008) provides a tool to study predicted miRNA targets and miRNA expression in human, mouse and rat. Target predictions are acquired from the miRanda algorithm and a more recent approach using a machine learning algorithm—miRSVR. Sequences for miRNA prediction from three mammalian species (human, mouse and rat) are acquired from mirBase and the UCSC genome browser. miRNA expression profiles are derived from next generation sequencing studies of human, mouse rat and cell lines from $\sim$250 small RNA libraries (Landgraf et al. 2007).

3.4.3 MicroCosm

Microcosm (http://www.ebi.ac.uk/enright-srv/microcosm/htdocs/targets/v5/) is a regularly updated database of predicted targets from miRanda provided by EMBL-EBI. MicroCosm v5 contains target predictions for mirBase sequences against the latest builds for EnsEMBL across 22 species' genomes. The user can search for interaction via miRNA ID, keyword, EnsEMBL ID or genome ontology (GO) term. Results outputted include the miRanda score, thermodynamic stability and the statistical significance of the predicted interaction.

3.4.4 *Tarbase*

Tarbase (Vergoulis et al. 2011) is a collection of supporting experimental data for miRNA-mRNA targets covering nine organisms from a range of experimental techniques including luciferase reporter assays, high throughput sequencing and microarrays. Version 6 of the database contains in excess of 65,000 thousand entries generated by initially scanning the literature using a text-mining pipeline followed by manual curation of parsed entries. Each entry contains the type of experimental technique from which the interaction was drawn, manuscript citation, alignment of the seed region and biological relevance. Tarbase is also cross linked with other databases including EnsEMBL and Uniprot.

3.4.5 *MirTarBase*

MirTarBase (Hsu et al. 2011) is a database of 4,270 experimentally confirmed targets for 14 species. The current build of the database (Version 2.5) covers 569 miRNAS and 2,533 targets extracted from a total of 1,380 articles in a similar manner to Tarbase (automatic literature mining followed by manual curation). Experimental data is predominantly from reporter construct assays or Western blotting and to a lesser degree from SILAC based proteomic and microarray expression profiling. The user can conduct a species specific search for the miRNA of interest to return target interactions. Each target interaction result contains information on the miRNA, gene and article as well as links to complementary biological databases.

3.4.6 *StarBase*

Recent years have seen increased application of next-generation sequencing technologies such as HTS-clip and degradome sequencing for analysis of miRNA mechanisms through Argonaute protein binding and transcriptional degradation. The starBase (Yang et al. 2011) database was developed to store the information arising from such experiments and to integrate these complex data with complementary resources. The current build of the database contains thousands of miRNA-mRNA interactions as well as information on Ago and other RNA binding proteins across six organisms. The database contains a variety of tools including a genome browser, and a search system to find interactions and integrate data with biological pathway information such as KEGG, Biocarta and GO.

3.4.7 miRecords

The miRecords (Xiao et al. 2009) database contains both validated and predicted animal miRNA targets. Predicted targets stored within miRecords are acquired from a number of target prediction algorithms including TargetScan, PicTar, PITA, RNAhybrid, DIANA-microT and RNA22 as well as some machine learning approaches. Also included are manually curated entries for validated targets from several species extracted from the literature. At present the database holds 2,286 interactions with experimental evidence (677 arise from profiling studies). From a search result a user can see the degree of overlap between predicted target algorithms along with the results from the literature search including experimental manipulation of the miRNA (over or under-expression) and effect on the mRNA or protein level.

3.4.8 miR2disease

As its name suggests miR2disease (Jiang et al. 2009) focuses on cataloguing those miRNA with an impact on human disease. The database contains 3,273 entries for 349 miRNA disregulated in 163 diseases, users can search for miRNAs, genes or a disease of interest. For example a search for stomach cancer returns a list of miRNAs with differential expression for several studies along with their role within the disorder (e.g. causal). Each search returns the method used to detect miRNA expression level, and validated targets from within the study or TarBase and links to predicted targets for the microRNA.

3.5 miRNA Prediction Algorithm Performance

Which algorithm is most likely to provide direct miRNA targets? At present the question is difficult to answer accurately due to large disparity between the numbers of confirmed targets in comparison to the number of predicted targets. For example, one study (Sethupathy et al. 2006) compared target predictions for 84 mammalian targets with 32 miRNAs in the current version of Tarbase that were confirmed using direct methods, e.g. 3′ luciferase assays. The study suffers from an obvious lack of experimentally confirmed targets given the breadth of miRNA regulation; however it is interesting to compare the performance of the individual algorithms against this set. To determine the false positive rate, a set of $\sim$20 targets confirmed to have no interaction were used. These experimentally confirmed targets and non-targets were compared against those predicted by TargetScan, TargetScanS, DIANA-microT, miRanda and PicTar along with the unions and intersections of various combinations of the outputs of these algorithms. Each set of predictions were compared in terms of sensitivity (S) (probability of detecting a true interaction). Early algorithms [TargetScan (S = 21 %), DIANA-microT (S = 10 %)] were found to have low sensitivity, with more recent algorithms [(miRanda (S = 49 %), TargetScanS (S = 48 %) and PicTar (S = 48 %)] significantly increasing in sensitivity. Combining the results from the five algorithms had the highest sensitivity (S = 100 %) outperforming individual algorithms and also the union of predictions.

Proteomic and transcriptomic profiling following miRNA transfection/knockout have also been utilised to test prediction algorithms. Two studies compared predicted outputs from miRBase Targets, miRanda, PicTar, PITA and TargetScan in terms of those principles outlined in Sect. 3.2 of this chapter. In this comparison PicTar and TargetScan had the highest prediction rate for those proteins repressed amongst the algorithms (although > 66 % of the predicted targets were not detected). The authors noted that this was due to the utilisation of site conservation and perhaps stringent pairing with the seed—analysis of perfect seed match against mismatched seed reduced the benefit of conservation analysis. Ranking predictions based on total context score (Grimson et al. 2007) proved to be a particularly effective means to identify proteins that were downregulated (Baek et al. 2008). Further analysis of the precision (P) (P = correct prediction/total prediction) and sensitivity of in silico target predictions (Alexiou et al. 2009) on the proteome (Selbach et al. 2008) further highlighted the effectiveness of TargetScan/TargetScanS (P = 51 %/P = 49 %) and PicTar(P = 49 %). In comparison, the miRanda (P = 29 %), PITA (P = 26 %) and RNA22 (P = 24 %) algorithms were found to have a low precision with high false positive rates.

It is clear that there is no one target prediction algorithm detects all targets of a miRNA. Current consensus when prioritising targets is to utilise multiple target prediction algorithms (Table 3.1) in combination with sequence and laboratory analysis. The algorithms chosen should focus both on seed site conservation (e.g. TargetScan, PicTar and Diana-microT) and thermodynamic calculations (PITA, RNA22). If possible global profiling tools such as proteomics or microarrays should be used to compare expression patterns of the predicted targets. In addition the sequence features of targets such as multiple sites in close proximity should be used to further identify candidates (Table 3.2).

3.6 Utilisation of miRNA Target Prediction in CHO Research

The CHO field is beginning to move beyond sequence and differential expression analysis to determining miRNA function. Two recent studies have illustrated how the guidelines listed above and integration of predictions with analytical data can be applied for candidate selection. The first study investigated miRNA expression and gene expression in parallel across the growth cycle (lag, exponential, stationary and decline) in order to determine the effect of miRNA on CHO cell function and transcriptome (Bort et al. 2011). Clusters of genes with similar expression patterns were compared to the target predictions for differentially expressed miRNAs from the 11 target prediction algorithms via the GeneSet2miRNA interface (Antonov et al. 2009). Only targets predicted by at least four algorithms were retained. Anti-correlation of

Table 3.1 Summary of target prediction algorithms. A brief summary of the main features of each algorithm is provided along with an accompanying website address

Algorithm	Method	Conservation	Website	Data retrieval
miRanda	Targets scored using sequence alignment, site location within UTR thermodynamics	Y	http://www.microrna.org	Search by transcript or miRNA
PicTar	Predict targets for multiple miRNAs (coexpression), targets scored on complementarity, free energy threshold, calculation of probabilities using HMM	Y	http://pictar.mdc-berlin.de/	Search by transcript or miRNA
Diana-microT	miRNA driver sequence complementarity, free energy binding applied to targets with perfect complementarity	Y	http://diana.cslab.ece.ntua.gr/microT/	Search by transcript or miRNA
TargetScan	Context scoring based on seed match, thermodynamic stability, AU composition	Y	http://www.targetscan.org/	Search by transcript or miRNA
TargetScanS	6 nt seed match with adenosine at miRNA position 1 + calculation of context score	Y	http://www.targetscan.org/	Search by transcript or miRNA
PITA	Determines free energy of target accessibility of target along with duplex formation energy	Y	http://genie.weizmann.ac.il/pubs/mir07/mir07_prediction.html	Search by transcript or miRNA or sequence
RNAhybrid	Identifies targets using thermodynamic calculations requiring matches to the seed region. Calculates significance of target using a shuffled sequence	Y	http://bibiserv.techfak.uni-bielefeld.de/rnahybrid/	Download program Windows, Mac or Linux
RNA22	Species specific identification of targets via alignment of variable length motifs and thermodynamic assessment of "hotspots"	N	http://cbcsrv.watson.ibm.com/rna22.html	miRNA and mRNA Sequence search
mirWalk	Target identification across mRNA, initial identification of seed binding and extension of alignment until mismatch located	Y	http://www.umm.uni-heidelberg.de/apps/zmf/mirwalk/index.html	Search by transcript or miRNA
HOCTAR	Target identification through combination of gene expression analysis and traditional target prediction algorithms	Indirect through overlap with other algorithms	http://hoctar.tigem.it/index.php	Search by transcript or miRNA

Table 3.2 Summary of online data repositories for microRNAs. The nature of the data included, source of miRNA target predictions (if present) and web address of resources are presented below

Database	Data type	Prediction algorithm	Analytical techniques	URL
miRBase	Sequence	NA	NA	http://www.mirbase.org/
microRNA.org	Computational prediction	miRanda miRSVR	NA	www.microRNA.org
microCosm	Computational prediction	miRanda	NA	http://www.ebi.ac.uk/enright-srv/microcosm/htdocs/targets/v5/
Tarbase	Confirmed targets Expression profiling	NA	Luciferase, Microarray, NGS, Proteomics	http://www.microrna.gr/tarbase
MirTarbase	Construct validation Expression profiling data	NA	Microarray, NGS, Proteomics	http://mirtarbase.mbc.nctu.edu.tw/
starBase	microRNA-target interactions, protein-RNA interaction, Degradome sequences	NA	HTS-CLIP, Degradome-Seq	http://starbase.sysu.edu.cn/
miRecords	Computational and Confirmed Targets	TargetScan, PicTar, PITA, RNAhybrid, DIANA-microT, RNA22	Luciferase, Microarray, NGS, Proteomics	http://mirecords.biolead.org/doc.php

miRNA and targets gene expression at specific phases of culture was determined to further prioritise targets and enrichment of biological processes relating to the stages growth rate (e.g. cell proliferation, cell cycle and apoptosis) was observed for groups of targets inversely related to their corresponding miRNAs.

Proteomic profiling has also been carried out in CHO cells with artificially overexpressed miR7 (Meleady et al. 2012), known to inhibit CHO cell growth and increase productivity (Barron et al. 2011). Quantitative label-free mass spectrometry was utilised to measure the differences between negative control cells and cells transfected with miR7 pre-mirs, with a total of 93 and 74 proteins found to be downregulated and upregulated respectively. Potential targets of those downregulated proteins were identified through the miRWalk interface by combining the outputs of six target prediction algorithms (miRanda, miRwalk, miRDB, RNA22, RNAhybrid and TargetScan). The miRWalk system was configured to identify target matches with a minimum seed length of seven in the 5′UTR, 3′UTR, promoter or CDS. Only targets with a p-value of < 0.05 were returned and targets were deemed significant if they were identified by two or more algorithms in either human, mouse or rat. The analysis identified catalase (CAT) and stathmin (STMN1) as potential direct targets of miR7

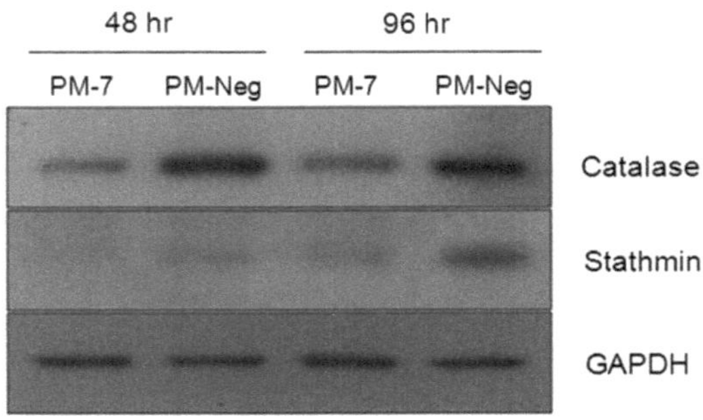

Fig. 3.2 Western blot analysis of catalase and stathmin 48 and 96 hrs after miR7 transfection. Both proteins were identified as potential targets following analysis with several in-silico target prediction methods. (Adapted from Meleady et al. 2012)

through overlap with several algorithm predictions in mouse and rat. Shown below (Fig. 3.2) is western blot analysis of these proteins following transfection with miR7 and the negative control, where the expression levels of both proteins are reduced in comparison to cell with negative control cells at 48 and 96 h prioritising these proteins for further experimental confirmation.

Expression profiling can aid in prioritising candidates for follow-up, however expression level alone of a gene/protein changing cannot be directly attributed to post-transcriptional regulation by a particular miRNA. Emergent techniques such as HTS-CLIP which identify miRNA-mRNA interaction on a global scale (Chi et al. 2009) by direct analyses of sequences bound to the Ago proteins will increase the number of validated direct targets in the future allowing us to gain a much clearer picture of the performance of miRNA target prediction algorithm accuracy.

3.7 Conclusions

Target prediction algorithms will play a greater role in understanding CHO regulations as we attempt to understand the impact of miRNA regulation on industrial cell culture. The complexity of animal miRNA interactions has led to a proliferation of computational techniques designed to predict which transcripts are regulated by a particular miRNA. In this chapter a subset of the most widely used algorithms and data repositories are described. Traditional algorithms follow multiple general miRNA and mRNA sequence based rules, while emerging techniques utilise previously identified and experimental data to predict targets. The performance of these target prediction algorithms has not been completely quantified as of yet due to the lack of experimentally validated targets in the literature. There is no universal algorithm for all target types; a general set of best practice guidelines have been suggested and it is recommended to utilise several streams of evidence including algorithms that interrogate the seed region (alignment, conservation, the presence of multiple site etc.) and thermodynamic calculations and where possible, data from expression profiling studies should be incorporated to filter results.

Acknowledgements This work was supported by funding from Science Foundation Ireland (SFI) grant number 07/IN.1/B1323.

References

Alexiou P, Maragkakis M, Papadopoulos GL, Reczko M, Hatzigeorgiou AG (2009) Lost in translation: an assessment and perspective for computational microRNA target identification. Bioinformatics 25:3049–3055

Antonov AV, Dietmann S, Wong P, Lutter D, Mewes HW (2009) GeneSet2miRNA: finding the signature of cooperative miRNA activities in the gene lists. Nucleic Acids Res 37:323–328

Azzouzi I, Moest H, Winkler J, Fauchere JC, Gerber AP, Wollscheid B, Stoffel M, Schmugge M, Speer O (2011) MicroRNA-96 directly inhibits gamma-globin expression in human erythropoiesis. PLoS One 6:e22838

Baek D, Villen J, Shin C, Camargo FD, Gygi S P, Bartel DP (2008) The impact of microRNAs on protein output. Nature 455:64–71

Barron N, Kumar N, Sanchez N, Doolan P, Clarke C, Meleady P, O'Sullivan F, Clynes M (2011) Engineering CHO cell growth and recombinant protein productivity by overexpression of miR-7. J Biotechnol 151:204–211

Bartel DP (2009) MicroRNAs: target recognition and regulatory functions. Cell 136:215–233

Betel D, Wilson M, Gabow A, Marks DS, Sander C (2008) The microRNA.org resource: targets and expression. Nucleic Acids Res 36:D149–153

Bort JA, Hackl M, Hoflmayer H, Jadhav V, Harreither E, Kumar N, Ernst W, Grillari J, Borth N (2011) Dynamic mRNA and miRNA profiling of CHO-K1 suspension cell cultures. Biotechnol J

Brennecke J, Stark A, Russell RB, Cohen SM (2005) Principles of microRNA-target recognition. PLoS Biol 3:e85

Chi SW, Zang JB, Mele A, Darnell RB (2009) Argonaute HITS-CLIP decodes microRNA-mRNA interaction maps. Nature 460:479–486

Doench JG, Sharp PA (2004) Specificity of microRNA target selection in translational repression. Genes Dev 18:504–511

Druz A, Chu C, Majors B, Santuary R, Betenbaugh M, Shiloach J (2011) A novel microRNA mmu-miR-466h affects apoptosis regulation in mammalian cells. Biotechnol Bioeng 108:1651–1661

Dweep H, Sticht C, Pandey P, Gretz N (2011) miRWalk–database: prediction of possible miRNA binding sites by "walking" the genes of three genomes. J Biomed Inform 44:839–847

Enright AJ, John B, Gaul U, Tuschl T, Sander C, Marks DS (2003) MicroRNA targets in Drosophila. Genome Biol 5:R1

Farh KK, Grimson A, Jan C, Lewis BP, Johnston WK, Lim LP, Burge CB, Bartel DP (2005) The widespread impact of mammalian MicroRNAs on mRNA repression and evolution. Science 310:1817–1821

Filipowicz W, Bhattacharyya SN, Sonenberg N (2008) Mechanisms of post-transcriptional regulation by microRNAs: are the answers in sight? Nat Rev Genet 9:102–114

Friedman RC, Farh KK, Burge CB, Bartel DP (2009) Most mammalian mRNAs are conserved targets of microRNAs. Genome Res 19:92–105

Gammell P, Barron N, Kumar N, Clynes M (2007) Initial identification of low temperature and culture stage induction of miRNA expression in suspension CHO-K1 cells. J Biotechnol 130:213–218

Garcia DM, Baek D, Shin C, Bell GW, Grimson A, Bartel DP (2011) Weak seed-pairing stability and high target-site abundance decrease the proficiency of lsy-6 and other microRNAs. Nat Struct Mol Biol 18:1139–1146

Gennarino VA, Sardiello M, Avellino R, Meola N, Maselli V, Anand S, Cutillo L, Ballabio A, Banfi S (2009) MicroRNA target prediction by expression analysis of host genes. Genome Res 19:481–490

Gennarino VA, Sardiello M, Mutarelli M, Dharmalingam G, Maselli V, Lago G, Banfi S (2011) HOCTAR database: a unique resource for microRNA target prediction. Gene 480:51–58

Grimson A, Farh KK, Johnston WK, Garrett-Engele P, Lim LP, Bartel DP (2007) MicroRNA targeting specificity in mammals: determinants beyond seed pairing. Mol Cell 27:91–105

Hackl M, Jakobi T, Blom J, Doppmeier D, Brinkrolf K, Szczepanowski R, Bernhart SH, Siederdissen CH, Bort JA, Wieser M, Kunert R, Jeffs S, Hofacker IL, Goesmann A, Puhler A, Borth N, Grillari J (2011) Next-generation sequencing of the Chinese hamster ovary microRNA transcriptome: Identification, annotation and profiling of microRNAs as targets for cellular engineering. J Biotechnol 153:62–75

Hammond S, Swanberg JC, Polson SW, Lee KH (2011) Profiling conserved MicroRNA expression in recombinant CHO cell lines using Illumina sequencing. Biotechnol Bioeng

Hobert O (2004) Common logic of transcription factor and microRNA action. Trends Biochem Sci 29:462–468

Hofacker IL (2003) Vienna RNA secondary structure server. Nucleic Acids Res 31:3429–3431

Hsu SD, Lin FM, Wu WY, Liang C, Huang WC, Chan WL, Tsai WT, Chen GZ, Lee CJ, Chiu CM, Chien CH, Wu MC, Huang CY, Tsou AP, Huang HD (2011) miRTarBase: a database curates experimentally validated microRNA-target interactions. Nucleic Acids Res 39:D163–169

Hughes AE, Bradley DT, Campbell M, Lechner J, Dash DP, Simpson DA, Willoughby CE (2011) Mutation altering the miR-184 seed region causes familial keratoconus with cataract. Am J Hum Genet 89:628–633

Jiang P, Wu H, Wang W, Ma W, Sun X, Lu Z (2007) MiPred: classification of real and pseudo microRNA precursors using random forest prediction model with combined features. Nucleic Acids Res 35:W339–344

Jiang Q, Wang Y, Hao Y, Juan L, Teng M, Zhang X, Li M, Wang G, Liu Y (2009) miR2Disease: a manually curated database for microRNA deregulation in human disease. Nucleic Acids Res 37:D98–104

John B, Enright AJ, Aravin A, Tuschl T, Sander C, Marks DS (2004) Human MicroRNA targets. PLoS Biol 2:e363

Johnson KC, Jacob NM, Nissom PM, Hackl M, Lee LH, Yap M, Hu WS (2011) Conserved microRNAs in Chinese hamster ovary cell lines. Biotechnol Bioeng 108:475–480

Kertesz M, Iovino N, Unnerstall U, Gaul U, Segal E (2007) The role of site accessibility in microRNA target recognition. Nat Genet 39:1278–1284

Kim SK, Nam JW, Rhee JK, Lee WJ, Zhang BT (2006) miTarget: microRNA target gene prediction using a support vector machine. BMC Bioinformatics 7:411

Kozomara A, Griffiths-Jones S (2011) miRBase: integrating microRNA annotation and deep-sequencing data. Nucleic Acids Res 39:D152–157

Krek A, Grun D, Poy MN, Wolf R, Rosenberg L, Epstein EJ, MacMenamin P, da Piedade I, Gunsalus KC, Stoffel M, Rajewsky N (2005) Combinatorial microRNA target predictions. Nat Genet 37:495–500

Kruger J, Rehmsmeier M (2006) RNAhybrid: microRNA target prediction easy, fast and flexible. Nucleic Acids Res 34:W451–454

Lal A, Navarro F, Maher CA, Maliszewski LE, Yan N, O'Day E, Chowdhury D, Dykxhoorn DM, Tsai P, Hofmann O, Becker KG, Gorospe M, Hide W, Lieberman J (2009) miR-24 Inhibits cell proliferation by targeting E2F2, MYC, and other cell-cycle genes via binding to "seedless" 3′UTR microRNA recognition elements. Mol Cell 35:610–625

Landgraf P, Rusu M, Sheridan R, Sewer A, Iovino N, Aravin A, Pfeffer S, Rice A, Kamphorst AO, Landthaler M, Lin C, Socci ND, Hermida L, Fulci V, Chiaretti S, Foa R, Schliwka J, Fuchs U, Novosel A, Muller RU, Schermer B, Bissels U, Inman J, Phan Q, Chien M, Weir DB, Choksi R, De Vita G, Frezzetti D, Trompeter HI, Hornung V, Teng G, Hartmann G, Palkovits M, Di Lauro R, Wernet P, Macino G, Rogler CE, Nagle JW, Ju J, Papavasiliou FN, Benzing T, Lichter P, Tam W, Brownstein MJ, Bosio A, Borkhardt A, Russo JJ, Sander C, Zavolan M, Tuschl T (2007) A mammalian microRNA expression atlas based on small RNA library sequencing. Cell 129:1401–1414

Lee I, Ajay SS, Yook JI, Kim HS, Hong SH, Kim NH, Dhanasekaran SM, Chinnaiyan AM, Athey BD (2009) New class of microRNA targets containing simultaneous 5′-UTR and 3′-UTR interaction sites. Genome Res 19:1175–1183

Lee RC, Feinbaum RL, Ambros V (1993) The C. elegans heterochronic gene lin-4 encodes small RNAs with antisense complementarity to lin-14. Cell 75:843–854

Lewis BP, Burge CB, Bartel DP (2005) Conserved seed pairing, often flanked by adenosines, indicates that thousands of human genes are microRNA targets. Cell 120:15–20

Lewis BP, Shih IH, Jones-Rhoades MW, Bartel DP, Burge CB (2003) Prediction of mammalian microRNA targets. Cell 115:787–798

Lim LP, Lau NC, Garrett-Engele P, Grimson A, Schelter JM, Castle J, Bartel DP, Linsley PS, Johnson JM (2005) Microarray analysis shows that some microRNAs downregulate large numbers of target mRNAs. Nature 433:769–773

Lin N, Davis A, Bahr S, Borgschulte T, Achtien K, Kayser K (2010) Profiling highly conserved microrna expression in recombinant IgG-producing and parental chinese hamster ovary cells. Biotechnol Prog

Lytle JR, Yario TA, Steitz JA (2007) Target mRNAs are repressed as efficiently by microRNA-binding sites in the 5′ UTR as in the 3′ UTR. Proc Natl Acad Sci U S A 104:9667–9672

Maragkakis M, Alexiou P, Papadopoulos GL, Reczko M, Dalamagas T, Giannopoulos G, Goumas G, Koukis E, Kourtis K, Simossis VA, Sethupathy P, Vergoulis T, Koziris N, Sellis T, Tsanakas P, Hatzigeorgiou AG (2009a) Accurate microRNA target prediction correlates with protein repression levels. BMC Bioinformatics 10:295

Maragkakis M, Reczko M, Simossis VA, Alexiou P, Papadopoulos GL, Dalamagas T, Giannopoulos G, Goumas G, Koukis E, Kourtis K, Vergoulis T, Koziris N, Sellis T, Tsanakas P, Hatzigeorgiou AG (2009b) DIANA-microT web server: elucidating microRNA functions through target prediction. Nucleic Acids Res 37:W273–276

Meleady P, Gallagher M, Clarke C, Henry M, Sanchez N, Barron N, Clynes M (2012) Impact of miR-7 over-expression on the proteome of Chinese hamster ovary cells. J Biotechnol

Mencia A, Modamio-Hoybjor S, Redshaw N, Morin M, Mayo-Merino F, Olavarrieta L, Aguirre LA, del Castillo I, Steel KP, Dalmay T, Moreno F, Moreno-Pelayo MA (2009) Mutations in the seed region of human miR-96 are responsible for nonsyndromic progressive hearing loss. Nat Genet 41:609–613

Miranda KC, Huynh T, Tay Y, Ang YS, Tam WL, Thomson AM, Lim B, Rigoutsos I (2006) A pattern-based method for the identification of MicroRNA binding sites and their corresponding heteroduplexes. Cell 126:1203–1217

Moretti F, Thermann R, Hentze MW (2010) Mechanism of translational regulation by miR-2 from sites in the 5′ untranslated region or the open reading frame. Rna 16:2493–2502

Nielsen CB, Shomron N, Sandberg R, Hornstein E, Kitzman J, Burge CB (2007) Determinants of targeting by endogenous and exogenous microRNAs and siRNAs. Rna 13:1894–1910

Orom UA, Lund AH (2010) Experimental identification of microRNA targets. Gene 451:1–5

Rehmsmeier M, Steffen P, Hochsmann M, Giegerich R (2004) Fast and effective prediction of microRNA/target duplexes. Rna 10:1507–1517

Reinhart BJ, Slack FJ, Basson M, Pasquinelli AE, Bettinger JC, Rougvie AE, Horvitz HR, Ruvkun G (2000) The 21-nucleotide let-7 RNA regulates developmental timing in Caenorhabditis elegans. Nature 403:901–906

Rhoades MW, Reinhart BJ, Lim LP, Burge CB, Bartel B, Bartel DP (2002) Prediction of plant microRNA targets. Cell 110:513–520

Saetrom P, Heale BS, Snove O Jr, Aagaard L, Alluin J, Rossi JJ (2007) Distance constraints between microRNA target sites dictate efficacy and cooperativity. Nucleic Acids Res 35:2333–2342

Schnall-Levin M, Rissland OS, Johnston WK, Perrimon N, Bartel DP, Berger B (2011) Unusually effective microRNA targeting within repeat-rich coding regions of mammalian mRNAs. Genome Res 21:1395–1403

Selbach M, Schwanhausser B, Thierfelder N, Fang Z, Khanin R, Rajewsky N (2008) Widespread changes in protein synthesis induced by microRNAs. Nature 455:58–63

Sethupathy P, Megraw M, Hatzigeorgiou AG (2006) A guide through present computational approaches for the identification of mammalian microRNA targets. Nat Methods 3:881–886

Sun G, Yan J, Noltner K, Feng J, Li H, Sarkis DA, Sommer SS, Rossi JJ (2009) SNPs in human miRNA genes affect biogenesis and function. Rna 15:1640–1651

Vergoulis T, Vlachos IS, Alexiou P, Georgakilas G, Maragkakis M, Reczko M, Gerangelos S, Koziris N, Dalamagas T, Hatzigeorgiou AG (2011) TarBase 6.0: capturing the exponential growth of miRNA targets with experimental support. Nucleic Acids Res 40:D222–D229

Watanabe Y, Yachie N, Numata K, Saito R, Kanai A, Tomita M (2006) Computational analysis of microRNA targets in Caenorhabditis elegans. Gene 365:2–10

Wu S, Huang S, Ding J, Zhao Y, Liang L, Liu T, Zhan R, He X (2010) Multiple microRNAs modulate p21Cip1/Waf1 expression by directly targeting its 3' untranslated region. Oncogene 29:2302–2308

Wuchty S, Fontana W, Hofacker IL, Schuster P (1999) Complete suboptimal folding of RNA and the stability of secondary structures. Biopolymers 49:145–165

Xiao F, Zuo Z, Cai G, Kang S, Gao X, Li T (2009) miRecords: an integrated resource for microRNA-target interactions. Nucleic Acids Res 37:D105–110

Xu X, Nagarajan H, Lewis NE, Pan S, Cai Z, Liu X, Chen W, Xie M, Wang W, Hammond S, Andersen MR, Neff N, Passarelli B, Koh W, Fan HC, Wang J, Gui Y, Lee KH, Betenbaugh MJ, Quake SR, Famili I, Palsson BO, Wang J (2011) The genomic sequence of the Chinese hamster ovary (CHO)-K1 cell line. Nat Biotechnol 29:735–741

Yan X, Chao T, Tu K, Zhang Y, Xie L, Gong Y, Yuan J, Qiang B, Peng X (2007) Improving the prediction of human microRNA target genes by using ensemble algorithm. FEBS Lett 581:1587–1593

Yang JH, Li JH, Shao P, Zhou H, Chen YQ, Qu LH (2011) starBase: a database for exploring microRNA-mRNA interaction maps from Argonaute CLIP-Seq and Degradome-Seq data. Nucleic Acids Res 39:D202–209

Yousef M, Jung S, Kossenkov AV, Showe LC, Showe MK (2007) Naive Bayes for microRNA target predictions—machine learning for microRNA targets. Bioinformatics 23:2987–2992

Chapter 4
The CHO miRNA Transcriptome

Matthias Hackl, Nicole Borth and Johannes Grillari

Abstract After the initial identification of microRNAs (miRNAs) almost 20 years ago (Lee et al, RNA 14(1): 35–42, 1993; Wightman et al, Cell 75(5): 855–862, 1993), research on the functional relevance of this class of small non-coding RNAs has increased exponentially, especially during the last 15 years. Today the importance of miRNAs as an additional layer in the regulation of gene expression is well appreciated and has established miRNAs as important research targets in virtually every area of cell biology from organism development (Alvarez-Garcia and Miska, Development 132(21): 4653–4662, 2005) to disease (Hammond, Curr Opin Genet Dev 16(1): 4–9, 2006) and cell death (Vecchione and Croce, Endocr-Rel Cancer 17(1): F37–50, 2010).

While several hundred miRNAs have been reported for human, murine and other mammalian species (Griffiths-Jones, Current Protocols in Bioinformatics 12.9.1–10, 2010), little attention has been attributed to miRNAs in the Chinese hamster, or the most widely used Chinese hamster derived cell line—Chinese hamster ovary (CHO) cells until recently. This is surprising, given the fact that miRNAs are known to orchestrate complex gene expression networks, thereby acting amidst important cellular networks that for example drive cell growth and survival and are consequently also of profound interest for the CHO research community, which is trying to adapt the cellular phenotype to the needs of modern bioprocesses.

Therefore, the aim of this chapter is to (i) provide a general overview on miRNA expression in mammalian cells; introduce the reader to potential applications of miRNAs in bioprocessing, (ii) to summarize the current knowledge about the miRNA transcriptome in CHO cells and (iii) to provide a perspective to the reader where this knowledge needs to be expanded so that future applications of miRNAs as bioprocessing tools will become feasible.

J. Grillari (✉) · M. Hackl · N. Borth
Department of Biotechnology, BOKU University Vienna, Muthgasse 18, 1190 Vienna, Austria
e-mail: johannes.grillari@boku.ac.at

N. Borth
ACIB Austrian Center of Industrial Biotechnology, Muthgasse 11, 1190 Vienna, Austria

N. Barron (ed.), *MicroRNAs as Tools in Biopharmaceutical Production*,
DOI 10.1007/978-94-007-5128-6_4, © Springer Science+Business Media Dordrecht 2012

4.1 MicroRNA Expression and Cellular Phenotypes in Mammalian Cells

Most of our knowledge on miRNA expression and functionality derives from studies performed in organisms ranging from *C. elegans* to human, with a notable lack of CHO data. Only a handful of studies have been published describing the use of transcriptomic tools for the analysis of microRNA (miRNA) expression in CHO cells (Table 4.1). Most of these studies were designed to measure miRNA levels under specific growth or culture conditions, in order to explore their functions and subsequently estimate their relevance for modulating CHO cell phenotypes. That changes in miRNA transcript levels can indeed impact on cellular phenotype seems realistic based on miRNA research in other mammalian organisms: in models of developmental biology, controlled changes in miRNA expression are known to mediate cell differentiation and organism development (Guo et al. 2011). In other studies co-delivery of hsa-miR-93 with transcription factors Oct-4, Klf-4, Sox2 and cMyc can increase the efficiency of re-differentiating somatic cells into pluripotent stem cells (Li et al. 2011). In addition the development of certain diseases was shown to be accompanied by changes in miRNA expression, for example as a consequence of genetic mutations or rearrangements. In this respect, specific miRNA expression signatures have been associated to various types of cancer (Lu 2005), since both oncogenes and tumor suppressor proteins are known to be post-transcriptionally controlled by miRNAs. Consequently, aberrant expression of miRNAs can allow for rapid and uncontrolled proliferation or loss of cell contact inhibition, which eventually will result in oncogenic cellular behavior (Iorio and Croce 2009).

Whether the biological relevance of miRNAs during these alterations in cell phenotypes lies in the induction or in the stabilization of the newly adopted cellular state, currently is unclear (Kosik 2010; Wu et al. 2009); that the transcription of miRNA genes and the subsequent processing of primary transcripts into mature miRNAs need to be highly regulated processes, which result in defined levels of mature miRNA transcripts depending on cell type or environmental condition is undisputed (Krol et al. 2010). At transcriptional level, this control is mediated by RNA polymerase II (Pol II) dependent transcription of most miRNA genes (Fig. 4.1), allowing Pol II-associated transcription factors to bind and regulate the production of primary miRNA transcripts (pri-miRs). In addition, the activity and levels of proteins involved in processing of pri-miRs into mature miRNAs—especially Drosha and Dgcr8 (Fig. 4.1)—affects the accumulation of miRNAs on a post-transcriptional level (Krol et al. 2010). Recently it was found that a number of well-known proteins, such as p53 or members of the SMAD family, are able to interact with Drosha and Dicer, thereby promoting the biogenesis of specific subsets of miRNAs (Suzuki et al. 2009; Davis et al. 2010). Overall these findings indicate that the composition of the miRNA transcriptome at a given timepoint under given conditions is not random but likely to significantly influence cellular state.

With respect to recombinant protein production cell lines, this means that measuring changes in miRNA levels in response to specific culture conditions or transitions in cellular state, is likely to increase our understanding of which genes and

Table 4.1 miRNA expression studies in Chinese hamster ovary (CHO) cells

CHO cell type	Cellular states/culture conditions	Differentially transcribed miRNAs	Detected miRNAs	Profiling method	Probe/primer design	Reference
CHO-K1	Temperature shift (72 h) (37 °C to 31 °C)	26	27.3 % present call	Microarray/qPCR	hsa, mmu	Gammell et al. 2007
CHO-K1	Temperature shift (24 h) (37 °C to 31 °C)	10	n/a	qPCR	hsa	Barron et al. 2011
CHO-S	Media nutrient depletion (depleted vs fresh media)	70	300	Microarray/qPCR	hsa, mmu, rno, miRBase 12.0	Druz et al. 2011
CHO-K1	Recombinant vs Host	5	54 (16)[a]	Microarray/qPCR	hsa, mmu, rno, miRBase 10.0	Lin et al. 2011
CHO-DG44	K1 vs DG44	5				
Not specified	Various CHO cell types	n/a	350[b]	Illumina NGS	Not applicable	Johnson et al. 2011
CHO-K1	Serum-free adapted vs serum-dependent	18	380[c]	Illumina NGS	Not applicable	Hackl et al. 2011
CHO-K1SV		8				
CHO-DUKXB-11	Recombinant vs Host					
CHO-K1	Growth phases during batch cultivation	118	199	Microarray/qPCR	hsa, mmu, rno, miRBase 9.2	Bort et al. 2011

Hsa *homo sapiens*, mmu *mus musculus*, rno *rattus norvegicus*, NGS next-generation sequencing, n/a not available

[a] 16 miRs selected from 54 candidate miRs with expression detected on microarray with high confidence

[b] Including similar isomiR sequences in the count

[c] Distinct miRNA sequences, not including isomiR sequences

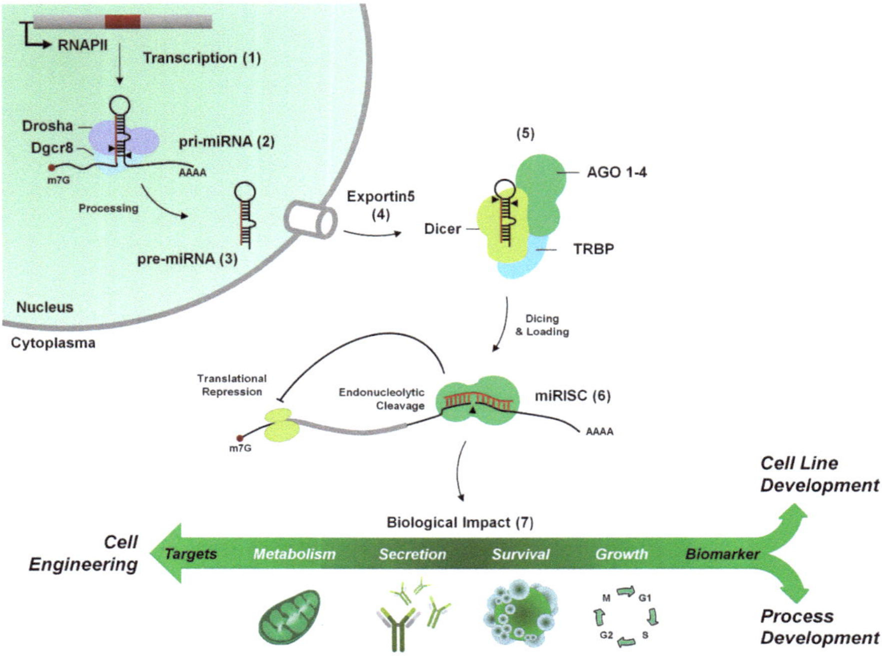

Fig. 4.1 miRNA expression in the context of biopharmaceutical production. Small non-coding RNAs referred to as microRNA (miRNAs) are transcribed from intergenic loci (1) or processed from intronic sequences of coding genes (not shown). After transcription of primary miRNA transcripts (pri-miRs, 2) from miRNA genes in an RNA Polymerase II (RNAPII) dependent manner, pri-miRs are processed into a ~ 70 base long precursor miRNA (pre-miR, 3) that becomes exported into the cytoplasm via Exportin 5 (4). In the cytoplasm the pre-miR gets incorporated into a protein complex consisting of RNaseIII enzyme Dicer, the RNA binding protein TRBP as well as members of the AGO protein family (5). Cleavage of the miRNA loop (*dicing*) results in the formation of the RNA-induced silencing complex (RISC) with the mature 22 base-long miRNA in its core (6). Depending on partially complementary sequence matches between a miRNA and its mRNA targets, the RISC is directed to an mRNA molecule—most frequently its 3′ untranslated regions (UTR)—resulting in either post-transcriptional repression or degradation (6). By these means, miRNAs become important regulatory molecules amidst many important cellular pathways (7), which in the context of biopharmaceutical production suggest an application of miRNAs as targets for cell line engineering and as biomarkers in cell line development and bioprocess development

miRNAs are related to specific phenotypes. This will further enable the development of miRNA-based tools both for better prediction and control of CHO cell behavior under bioprocessing relevant conditions.

4.2 MicroRNAs in the Context of Biopharmaceutical Production: EngimiRs and Biomarkers

Two future applications of miRNAs in the context of biopharmaceutical production are currently envisioned as most promising by the research community (Fig. 4.1): first, the use of miRNAs as targets for cell engineering approaches (engimiRs) through which the performance of cell factories for recombinant protein production can be enhanced; second, the use of miRNAs as indicators of CHO cell phenotypes, eventually resulting in their application as predictors of specific cellular traits such as productivity during cell line development—as been recently achieved using mRNA levels (Clarke et al. 2011)—or as indicators of cellular state during process optimization.

In the light of published studies in other species, both applications seem realistic: genetic mutations, deletions or heterologous expression of miRNAs in mammalian cell lines can severely impact cellular behavior in terms of growth (Yu et al. 2012), metabolism (Krützfeldt and Stoffel 2006), and protein translation (Nottrott et al. 2006), as well as stress resistance and cell death (Frank et al. 2011). This was shown for several miRNAs, but most remarkably in the case of miR-21. Initially miR-21 drew attention as it was repeatedly found overexpressed in many different tumour types. In vitro studies that followed these observations attributed miR-21 oncogenic activity by showing that it negatively regulates the expression of numerous tumor suppressor protein mRNAs (Zhu et al. 2008). Finally, also in vivo miR-21 was confirmed as an oncogene (*oncomiR*) by Frank Slack's lab, showing that conditional overexpression of miR-21 in mice results in the formation of malignant B-Lymphocytes. In contrast, inactivation of miR-21 overexpression caused the tumors to regress, presumably due to increased levels of apoptosis (Medina et al. 2010). Similarly, other miRNAs have been implicated in the regulation of glucose/glutamine metabolism (Zhu et al. 2011; Gao et al. 2009) and cell cycle regulation (Cloonan et al. 2008), p53 mediated stress response (Hermeking 2010), or protein secretion (Lovis et al. 2008). For an in-depth discussion of putative engineering candidates please consult Chap. 5 of this book and respective reviews (Barron et al. 2011; Müller et al. 2008).

The possibility of applying miRNAs as biomarkers of cellular state during cell line development and process optimization, derives from technologies that are being developed for prognosis and diagnosis of diseases: by mining the plentitude of miRNA transcriptome data generated from several disease specimens, the levels of single or sets of miRNAs (so-called expression signatures) can be correlated to the presence or absence of a disease or even to determine the disease state for a given patient (Jeffrey 2008). Initally RNA extracts for biomarker discovery were obtained from tissue samples. Recently, researchers also discovered that miRNAs along with other types of RNAs and proteins are secreted from cells into the bloodstream (or find their way into the bloodstream after lysis of diseased cells). The field has now been exteneded to plasma and serum biomarker discovery (Gilad et al. 2007). Potential analogous bioprocessing applications lie in the identification of intracellular or extracellular miRNAs that can indicate early on whether a cell harbors a favorable phenotype or not, thus shortening the timelines during cell line development and making screening more targeted and efficient.

Although the road towards these future applications is paved by innovations in the field of diagnosis and treatment of disease, what is missing is a profound characterization of miRNA expression in common recombinant protein production cell lines, including both CHO, human (HEK293, PerC6) or other animal derived cell lines; and while few CHO miRNA profiling studies have already been undertaken using cross-species array platforms (Table 4.1), the major factor contributing to the lack of CHO specific miRNA expression data—the absence in public miRNA sequence data—has only recently been overcome.

4.3 Breaking the Ground for CHO-Specific miRNA Profiling

The often cited lack of availability of public information on genomic and transcriptomic sequences derived from CHO cells has resulted in an expanded use of cross-species platforms for measuring RNA transcript levels. While this requires sophisticated verification of results for mRNA array platforms (Yee et al. 2008; Ernst et al. 2006), the finding that miRNA sequences are generally well conserved across animals, with an even higher degree of conservation between human, mouse and rat (Berezikov 2005), has simplified this task for cross-species miRNA platforms. Consequently, all currently available CHO miRNA profiling data (Table 4.1) were acquired using either human, mouse or rat platforms except for two studies, which applied next-generation sequencing (RNA-seq) technologies (Hackl et al. 2011; Johnson et al. 2011).

Sequence data collected in these studies have been included in publications, and are likely to be incorporated into the miRBase miRNA sequence repository—a database harboring all currently known mature as well as precursor miRNA sequences—during its next update (Griffiths-Jones 2010). Since these studies were not primarily designed to pinpoint differentially expressed miRNAs, but to identify and annotate the maximum number of CHO miRNA sequences, a variety of different cell types and culture conditions were included in the sequencing efforts: sequencing libraries were generated from distinct CHO cell lines representing (i) adherent and suspension adapted cells, (ii) host and recombinant cells, (iii) cells cultivated in serum-free and serum-containing media and (iv) from a pool of RNA harvested from growth arrested CHO cells or after cold shock, heat shock or sodium butyrate treatment. By BLAST alignment or mapping sequencing reads to a reference composed of miRBase miRNA sequences, a high degree of miRNA sequence conservation between human, mouse, rat and Chinese hamster (*Cricetulus grisesus*, abbreviated with *cgr* in miRNA nomenclature) was confirmed: 350 conserved miRNAs (Johnson et al. 2011) compared to 380 conserved miRNAs (Hackl et al. 2011) were reported, which overlap by 340 miRNA sequences and therefore add up to a total of 390 currently identified mature miRNAs in CHO cells. Although the numbers of mature miRNAs reported for human (1,921), mouse (1,157) or rat (680) in miRBase version 18 are higher, tissue and cell type specific expression of miRNAs results in much lower numbers of expressed miRNAs in a cell line (Lee et al. 2008). Therefore, it is likely the set of 390 mature miRNAs reported in CHO constitutes the majority of highly conserved miRNAs that are expressed in CHO cells.

Based on the publication of the CHO-K1 genome in August 2011 (Xu et al. 2011), we also started to look at the genomic location of these 390 CHO miRNAs in the given reference genome (Hackl et al. 2012): using BLAST alignment unambiguous

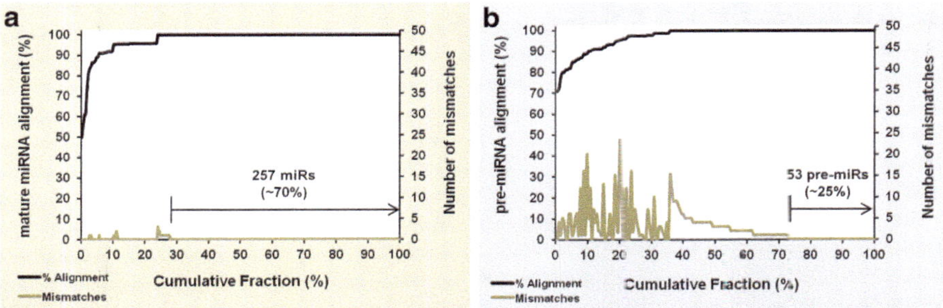

Fig. 4.2 *Conservation of CHO miRNAs.* 365 CHO miRNAs and the respective 212 precursor-miRNAs (pre-miRNAs) that gave an unambiguous BLAST alignment in the CHO-K1 genome were mapped against miRNAs from human, mouse and rat as available in the latest update of miRBase (version 18, Nov. 2011). **a** Out of 365 mature miRNAs, 257 (70.4 %) possess a 100 % identical (full-length alignment and no mismatches) human or rodent homolog. On the primary y-axis the fraction of the CHO miRNA in percent that could be aligned to a human or rodent ortholog is shown, while the secondary y-axis shows the number of mismatches observed for the respective alignment. **b** An analogous analysis for the 212 CHO pre-miRNAs identified 53 (25 %) sequences with perfectly identical human or rodent ortholog

hits for 365 mature miRNAs ($\sim$94 %) were observed, corresponding to 212 miRNA hairpins or miRNA gene loci. The respective sequences of both mature and precursor miRNAs were subsequently compared to their human, mouse and rat orthologs available in miRBase version 18 to find that 257 mature miRNAs (70.4 %) possess a 100 % identical (full-length alignment and no mismatches) human or rodent homolog (Fig. 4.2a). A similar analysis for precursor miRNA sequences identified 53 pre-miRNAs (25 % of CHO pre-miRNAs) in human, mouse or rat that are 100 % identical to the CHO ortholog (Fig. 4.2b). Thus, by the use of cross-species microarray platforms, which most commonly comprise combined probe-sets against human, mouse and rat miRNAs, more than 70 % of the currently annotated CHO miRNAs can be reliably detected.

Expansion of currently reported CHO miRNA sequences can be achieved by identification of novel and CHO-specific miRNAs, which was done by mapping small RNA reads to the mouse reference genome (Hackl et al. 2011). Initially 11 potential miRNA loci were thus identified, of which two miRNAs (cgr-miR-6091 and cgr-miR-6092) gave an unambiguous BLAST hit in the CHO-K1 reference genome, exhibited a canonical miRNA-like secondary structure and can therefore be considered as the first CHO-specific miRNAs.

To sum up, the currently available miRNA-seq data from CHO cells point to a high conservation of mature miRNA sequences compared to human and several rodent species. Consequently, this raises the question whether miRNA functions are equally well conserved. The presence of most (80 %) of the so far reported and validated miRNA binding sites for miR-17-92 in CHO cDNA sequence data suggests a similar biologic function at least for this miRNA cluster. However, comprehensvie studies of miRNA binding sites in CHO cDNA sequences have not yet been performed.

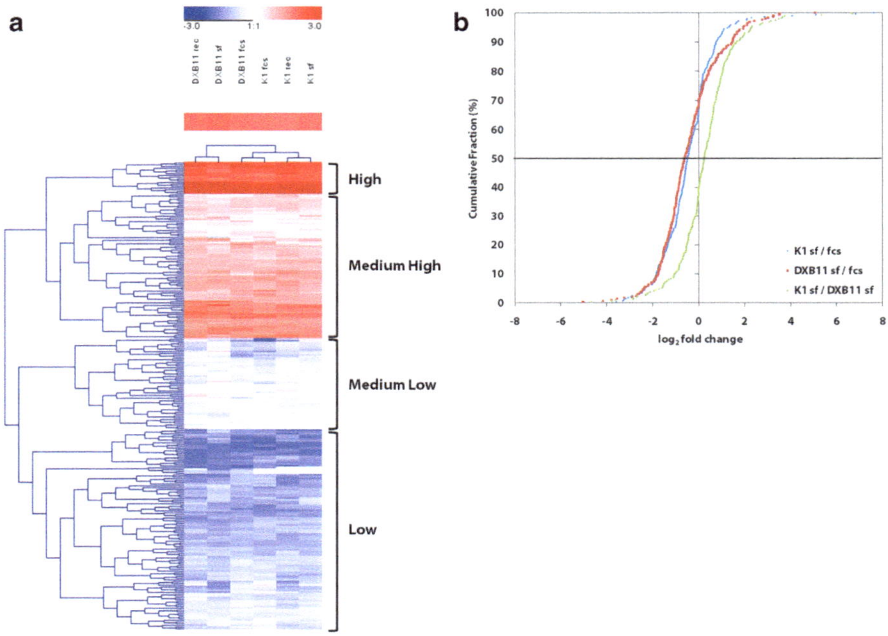

Fig. 4.3 *Transcription of miRNAs in CHO cell lines.* **a** Heatmap illustrating mature miRNA transcript levels in six distinct CHO cell lines, where clusters of high, medium and low abundant miRNAs are indicated. **b** Cumulative fraction plot showing $\log_2$ fold changes in miRNA expression for three different comparisons: CHO-K1 serum-free adapted (sf) versus serum-dependent (fcs), DXB11 sf versus fcs and CHO-K1 sf versus DXB11 sf. Results indicate a trend towards global down-regulation of miRNA transcription or processing in serum-free adapted CHO cell lines, but not for control comparison of serum-free CHO-K1 to DUKXB-11. (data derived from Hackl et al. 2011, ref. 18)

Therefore, several studies in CHO have focused to measure miRNA transcript levels under certain conditions, in order to relate their expression to cell type, proliferation, apoptosis and recombinant protein production (Table 4.1). In the following, a detailed summary of the respective results will be given.

4.4 The CHO miRNome

4.4.1 Correlation of miRNA Expression with Cellular State

Quantitative analysis of next-generation sequencing data showed that the majority of miRNAs in CHO cells exhibit low to medium expression levels (60 %) with only some miRNAs, especially members of the let-7 family and miR-21 (Hackl et al. 2012), that seem to be highly abundant in CHO cells (Fig. 4.3a). As indicated by the heatmap patterns, miRNA expression levels were generally observed to be similar between different CHO cell lines. However, explorative data analysis identified a clear difference between serum-free CHO-K1 cells and serum-free DUKXB-11 (dhfr$^{+/-}$) as well as in response to presence of serum in cultivation media (Hackl et al. 2011). Indeed, differential expression analysis showed that the adaptation of three distinct cell lines (CHO-K1, CHO-K1SV and DUKXB-11), both host and recombinant, to serum-free growth was accompanied by a global down-regulation of miRNA expression as indicated in a cumulative fraction plot calculated from available read count data (Hackl et al. 2011) (Fig. 4.3b). Given the fact that serum withdrawal from media commonly also decreases the specific growth rate, this observation would attribute miRNAs an important role in mediating growth of CHO cells, which shall be reviewed next.

4.4.2 Linking miRNA Expression and Cellular Proliferation in CHO Cells

A great amount of time and work has been devoted by the research community to elucidating the importance of miRNAs for cellular proliferation, mainly with the goal of gaining a better understanding of the oncogenic phenotype of tumor cells. Today, microarrays are extensively used to compare miRNA expression in various tumor cells or tissues to the respective normal controls, and miRNAs reoccurring in these screenings as up- or down-regulated in tumor cells are commonly designated as either oncomiRs or tumor suppressor miRNAs. To the dismay of cancer researchers, however, oncogenic or tumor suppressive functions of miRNAs turned out to be specific for certain types of cells or tissues (Lee and Dutta 2009) since miRNA function is likely to depend on the individual genetic background of cells (Brenner et al. 2010). For miRNA research in CHO cells this means that despite the availability of a variety of miRNA expression data from highly proliferating tumor cells, independent—CHO specific—data need to be generated, followed by functional characterization of prioritized miRNAs based on their expression characteristics.

In the first study on miRNA expression in CHO cell factories, Gammell et al. looked at changes in miRNA levels upon temperature shift from 37 °C to 31 °C (Gammell et al. 2007). This shift to lower temperature is commonly used in biphasic cultivation strategies to switch cells from a proliferative state to growth arrest, typically accompanied by prolonged viability as well as increased protein production and secretion (Trummer et al. 2006a, 2006b). While this initial study in 2007 was mainly focused on establishing and validating the chosen cross-species approach to miRNA profiling, in a subsequent study the cold-shock experiment was repeated, this time looking at acute temperature-sensitive miRNAs after 24 h incubation at 31 °C (Barron et al. 2011). Ten miRNAs with significant differential expression were detected, four down-regulated and six up-regulated, which by *in silico* prediction of target interactions were associated with regulation of nucleic acid metabolism and mRNA transcription, processes known to be reduced upon temperature reduction (Roobol et al. 2009). Among the down-regulated candidate miRNAs, miR-7 stood out with greater than 8-fold repression after temperature reduction. Due to its known interactions with mRNAs of Pak1, which is a member of p21 activating kinases, and insulin-growth factor receptor 1 (Jiang et al. 2010; Reddy et al. 2008), miR-7 was chosen for functional characterization by transient overexpression and knock-down. Results from these experiments attributed miR-7 a role as active inhibitor of cell growth upon overexpression, accompanied by an increase in recombinant protein production rates, thus mimicking the effects of biphasic cultivation strategies. Down-regulation of miR-7 using antisense oligonucleotides, however, did not result in increased growth of CHO cells suggesting redundant mechanisms in the gene expression control of its targets.

Another recent study focused in detail on the dynamics of miRNA levels during batch cultivation of CHO-K1 cells (Bort et al. 2011). Overall 118 miRNAs were identified that were significantly regulated during the course of a batch culture compared to lag phase by exhibiting a > 1.5 fold change in expression level over time. Cluster analysis of the expression profiles of these 118 candidates identified two principal groups: (i) 43 miRNAs (36 %) with constant increase in transcript levels towards stationary and decline phase, and (ii) 75 miRNAs (64 %) with peak levels during exponential phase and subsequent down-regulation during stationary and decline phase, therefore correlating with specific growth rate of cells during batch cultivation. Among the miRNAs sharing this profile, five distinct miRNA families (miRNAs with high sequence similarity) and six miRNA clusters (miRNAs organized in close genomic proximity and therefore belonging to the same transcriptional unit) were identified that were represented by at least two or more of its members.

The expression pattern of miRNAs of the miR-17-92 cluster was of particular interest, as this cluster has been previously implicated in the regulation of cell cycle progression (Cloonan et al. 2008; Pickering et al. 2009), proliferation (Manni et al. 2009) and oncogenicity (Olive et al. 2009). Several studies focusing on target identification ofmiR-17-92 indicate that this cluster mediates its effect by negative regulation of important cell cycle protein mRNAs such as CDKN1 A (p21), CDKN1 C (p57), or retinoblastoma 1 (Rb1), as well as apoptosis related protein mRNAs such as phosphatase tensin homolog (PTEN), or BCL2-like-11 (reviewd by Grillari et al. 2010). Based on RNA-seq data generated from different CHO cell lines (Becker et al. 2011), the majority of the reported target sites for miR-17-92 could be identified in CHO (Hackl et al. 2011), suggesting a similar function for miR-17-92 in CHO cells. Therefore, elevated levels of miR-17-92—as observed during exponential growth of CHO cells—could indirectly induce (or support) a proliferative state by suppressing cell cycle inhibitors p21 and p57, thus leading to an activation of E2 F transcription factor.

While this cluster of miRNAs was identified as a potential effector of CHO cell growth and survival by separate analysis of miRNA array-data, an integrated analysis of miRNA and mRNA microarray array data from CHO batch cultivation was also reported (Bort et al. 2011). The results indicate that in particular the up-regulation of a set of mRNAs during the late phases of batch cultivation where cell growth is slowing down to finally become arrested, may be enhanced or controlled by simultaneous down-regulation of a group of miRNAs, which preferentially bind to these mRNAs. This observed negative correlation in the expression of miRNAs and their predicted mRNA targets point towards functional miRNA~mRNA networks that regulate the growth of CHO cells.

4.4.3 ApoptomiRs—miRNAs in Control of Survival and Death of CHO Cells

Quite fittingly, the term *apoptomiR* was recently established to describe miRNAs that can influence death and survival of a cell by regulating cellular apoptosis (Vecchione and Croce 2010). This evolutionary conserved process, which enables an organism to remove cells no longer useful for its function or even dangerous for its survival was also shown to trigger the death of CHO cells in cultivation, for example in response to nutrient depletion or accumulation of toxic metabolites (Arden and Betenbaugh 2004). Hence, several cell engineering approaches employed over-expression of anti-apoptotic genes such as members of the BCL-2 superfamily of proteins, to protect CHO cells from undergoing apoptosis and consequently allow for higher cell densities and therefore space-time yields (Majors et al. 2008, 2009). Since it is likely that a combination of down-regulation and over-expression of specific pro- and anti-apoptotic genes is the most promising approach to control apoptosis in CHO cells, miRNAs, with their range of distinct targets and ability to stabilize cellular states, might prove to be valuable tools in this endeavor.

So far, only one study has specifically addressed the relevance of miRNAs for apoptosis in CHO cells by analyzing the changes in miRNA expression upon nutrient depletion, which was shown to increase Caspase-3/7 activity and consequently trigger apoptosis (Druz et al. 2011). Using microarray technology 70 miRNAs were identified as either significantly up or down-regulated in cells exposed to nutrient depleted media compared to fresh media controls. Among these, all 18 members of the miR-297-669 cluster that were detected on the array were found to be strongly induced in apoptotic versus control cells, with miR-466h being 452-fold induced in cells exposed to nutrient depleted media. Such a high level of induction, which is rarely observed for miRNAs, is possibly the result of miR-466h transcription or processing being shut-down under normal conditions, thus suggesting an important but also tightly controlled function for this miRNA: by *in silico* analyses of potential miR-466h targets, nine genes with anti-apoptotic function such as BCL2 or BCL2L2 were identified, thus pointing towards a pro-apoptotic function of miR-466h (Druz et al. 2011).

Druz et al. (2011) further went into functional characterization of miR-466h by again inducing apoptosis in CHO cells via nutrient depletion, but at the same time reducing miR-466h levels using antisense-inhibitor sequences. The suppression of miR-466h induction upon nutrient depletion improved the viability of CHO cells compared to negative controls and significantly delayed the onset of apoptosis, probably due to increased levels of several anti-apoptotic targets of miR-466h (Druz et al. 2011). Thus, miR-466h is the first cgr-apoptomiR that can decide over death and survival of CHO cells that has been identified.

4.4.4 miRNA Expression Signatures That Support Recombinant Protein Production

Besides enhanced growth and survival under stressful conditions, high recombinant protein productivity is an equally desirable trait of an industrially used cell line. With respect to CHO cells, media and process optimizations have lead to the most significant increases in recombinant protein titers without much increase in the specific protein production rates (qP) that commonly range between 25 and 50 pg/cell/day for monoclonal antibodies. However, since it is known that reduced temperature in biphasic bioprocesses or small molecules such as sodium butyrate can improve specific protein production rates, transcriptomics studies have been performed to identify gene expression changes under these conditions (Yee et al. 2008, 2009). While these studies gave valuable insights into the cellular pathways that might contribute or stabilize a cellular state of high recombinant protein productivity, they failed in pinning down individual candidates for engineering productivities.

Knowing that single miRNAs can be potent regulators of CHO cell growth and survival, the next step was to see whether miRNA expression signatures exist in CHO cells that can drive or support a state of high protein production and secretion. Lin et al. addressed this question with an initial comparison of miRNA expression between recombinant DG44 clones—amplified and non-amplified using methotrexate (MTX)—and the respective DG44 host cell line (Lin et al. 2010): based on the results from a cross-species miRNA microarray platform a panel of 16 miRNAs was selected for qPCR analyses in four IgG producing subclones as well as their respective parental cell line. Results of qPCR analysis showed that 8 out of 16 miRNAs exhibited a fold change > 1.5 in at least two out of four recombinant subclones, with two miRNAs—miR-221 and miR-222—being more than two-fold down-regulated and miR-17 being up-regulated in all four subclones.

To determine whether expression of these miRNAs correlated with recombinant IgG-productivities, MTX amplification was performed and changes in specific productivities (qP) were analyzed. In one of the clones that demonstrated a dose-dependent increase in qP with increasing MTX concentrations, the levels of miR-221/222 were analyzed at four stages during MTX amplification, however, no significant correlation between their expression and qP could be detected.

Nevertheless, in the light of an observation that was made based on miRNA-seq data from recombinant and host CHO cell lines (Hackl et al. 2011), the regulation of the mir-221 hairpin and its processing in respect to recombinant protein productivity is of interest: Dicer1 processing of miRNA hairpins (denoted as mirs) results in a short duplex sequence with $3'$ overhangs of which one strand is often (in $\sim 50\ \%$ of cases) rapidly degraded (the "star" strand, e.g. miR-221*) while the other strand is incorporated into a protein machinery and hence more stable and easier to detect (the mature strand, e.g. miR-221). The other possible outcome of miRNA processing is that a hairpin actually gives rise to two equally abundant mature miRNAs; in this case the affixes "-5p" and "-3p" are added to the names to distinguish both forms—a much clearer nomenclature, which will soon be installed for all miRNAs, thus replacing the often confusing mature/star nomenclature (personal communication with Dr. Griffiths-Jones).

In case of miRNAs originating from the mir-221 hairpin, miR-221-3p was observed in several species to be higher abundant than miR-221-5p, and therefore became known as miR-221, while miR-221-5p was further regarded as miR-221*. In CHO cells, however, not only equal expression levels for both mir-221 derived miRNAs were reported (Hackl et al. 2011), but also that the log-ratio in levels of miR-221-3p and miR-221-5p changes from positive (i.e. more miR-221–3p = miR-221) for both serum-free DHFR and K1 host cells to negative for the respective DHFR and K1 recombinant IgG producing subclones (i.e. less miR-221-3p). With this, two independent studies using different CHO cell types have identified a relationship between the regulation of miR-221 and recombinant cell phenotype. As functional characterization of mir-221 is most likely underway, it will be exciting to see whether targeted reduction of miR-221-3p levels lead to an improvement in recombinant protein production.

4.5 Perspectives on the Use of miRNA Tools in Biopharmaceutics

The data summarized here on miRNA expression and relevance in CHO cells clearly point to (i) a high number of expressed miRNAs, (ii) a dynamic in miRNA levels in response to culture conditions and cellular state and (iii) clear evidence that individual miRNAs like miR-7 and miR-466h can modify growth and recombinant protein production capacities or alter the stress responses of CHO cells, respectively. These first evidences for the opportunities of miRNA research in CHO cells also sets the need for developing further tools that will facilitate miRNA research in CHO cells: first, this encompasses the complete characterization of the miRNA transcriptome including the prediction and verification of novel (CHO-specific) small RNAs as well as large-scale analysis of miRNA transcription in various CHO cell lines. Secondly, the recent publication of the CHO mRNA transcriptome in the form 15,000 complete cDNA sequences with > 90 % coverage of homologous mouse transcripts (Becker et al. 2011) will allow a detailed analysis of miRNA target sites in CHO cells and whether they have evolved compared to its close rodent relatives or not. Based on such analyses as well as experimental approaches employing transcriptomics, proteomics and a combination of deep-sequencing and miRNA-target crosslinking (Chi et al. 2009; Nonne et al. 2010) it will be possible to gain insights into CHO-specific miRNA ~ mRNA target interactions.

Together, these tools will provide the basis to better understand individual or sets of miRNAs that can be used to modulate (engimiRs) or predict (biomarkers) CHO cell behavior in terms of proliferation and stress resistance as well as recombinant protein production and protein quality.

Acknowledgment MH would like to acknowledge support by the BOKU Doc doctoral scholarship program; NB and JG the support of the FWF Doctoral program "BioToP"; JG support by the GEN-AU project "Non-coding RNAs" (grant number 820982).

References

Alvarez-Garcia I, Miska EA (2005) MicroRNA functions in animal development and human disease. Development 132(21):4653–4662

Arden N, Betenbaugh MJ (2004) Life and death in mammalian cell culture: strategies for apoptosis inhibition. Trends Biotechnol 22(4):174–180

Barron N, Kumar N et al (2011) Engineering CHO cell growth and recombinant protein productivity by overexpression of miR-7. J Biotechnol 151(2):204–211

Barron Niall, Sanchez N et al (2011) MicroRNAs: tiny targets for engineering CHO cell phenotypes? Biotechnol Lett 33(1):11–21

Becker J et al (2011) Unraveling the Chinese hamster ovary cell line transcriptome by next-generation sequencing. J Biotechnol, accepted for publication

Berezikov E (2005) Phylogenetic shadowing and computational identification of human microRNA genes. Cell 120:21–24

Bort JAH et al (2011) Dynamic mRNA and miRNA profiling of CHO-K1 suspension cell cultures. Biotechnol J. Available at: http://www.ncbi.nlm.nih.gov/pubmed/21751394. Accessed August 15, 2011

Brenner JL et al (2010) Loss of individual microRNAs causes mutant phenotypes in sensitized genetic backgrounds in C. elegans. Curr Biol: CB, 20(14):1321–1325

Chi SW et al (2009) Argonaute HITS-CLIP decodes microRNA-mRNA interaction maps. Nature 460(7254):479–486

Clarke C et al (2011) Predicting cell-specific productivity from CHO gene expression. J Biotechnol 151(2):159–165

Cloonan N et al (2008) The miR-17-5p microRNA is a key regulator of the G1/S phase cell cycle transition. Genome Biol 9(8):R127

Davis BN et al (2010) Smad proteins bind a conserved RNA sequence to promote microRNA maturation by Drosha. Mol Cell 39(3):373–384

Druz A et al (2011) A novel microRNA mmu-miR-466h affects apoptosis regulation in mammalian cells. Biotechnol Bioeng 108(7):1651–1661

Ernst W et al (2006) Evaluation of a genomics platform for cross-species transcriptome analysis of recombinant CHO cells. Biotechnol J 1(6):639–650

Frank D et al (2011) MicroRNA-20a inhibits stress-induced cardiomyocyte apoptosis involving its novel target Egln3/PHD3. J Mol Cell Cardiol. Available at: http://www.ncbi.nlm.nih.gov/pubmed/22182733. Accessed February 21, 2012

Gammell P et al (2007) Initial identification of low temperature and culture stage induction of miRNA expression in suspension CHO-K1 cells. J Biotechnol 130(3):213–218

Gao P et al (2009) c-Myc suppression of miR-23a/b enhances mitochondrial glutaminase expression and glutamine metabolism. Nature 458(7239):762–765

Gilad S et al (2008) Serum MicroRNAs Are Promising Novel Biomarkers. PLoS ONE 3(9):e3148

Griffiths-Jones S (2010) miRBase: microRNA sequences and annotation. Curr Protoc Bioinformatics 12.9.1–10 (Editoral Board, Andreas D. Baxevanis... et al, Chapter 12)

Grillari J, Hackl M, Grillari-Voglauer R (2010) miR-17–92 cluster: ups and downs in cancer and aging. Biogerontology 11(4):501–506

Guo L, Zhao RCH, Wu Y (2011) The role of microRNAs in self-renewal and differentiation of mesenchymal stem cells. Exp Hematol 39(6):608–616

Hackl M et al (2011) Next-generation sequencing of the Chinese hamster ovary microRNA transcriptome: Identification, annotation and profiling of microRNAs as targets for cellular engineering. J Biotechnol 153(1–2):62–75

Hackl M et al (2012) Computational identification of microRNA gene loci and precursor microRNA sequences in CHO cell lines. J Biotechnol Available at: http://www.ncbi.nlm.nih.gov/pubmed/22306111. Accessed February 20, 2012

Hammond SM (2006) MicroRNAs as oncogenes. Curr Opin Genet Dev 16(1):4–9

Hermeking H (2010) The miR-34 family in cancer and apoptosis. Cell Death Differ 17(2):193–199

Iorio MV, Croce CM (2009) MicroRNAs in cancer: small molecules with a huge impact. J Clin Oncol 27(34):5848–5856

Jeffrey SS (2008) Cancer biomarker profiling with microRNAs. Nat Biotech 26(4):400–401

Jiang L et al (2010) MicroRNA-7 targets IGF1R (insulin-like growth factor 1 receptor) in tongue squamous cell carcinoma cells. Biochem J 432(1):199–205

Johnson KC et al (2011) Conserved microRNAs in Chinese hamster ovary cell lines. Biotechnol Bioeng 108(2):475–480

Kosik KS (2010) MicroRNAs and cellular phenotypy. Cell 143(1):21–26

Krol J, Loedige I, Filipowicz W (2010) The widespread regulation of microRNA biogenesis, function and decay. Nat Rev Genet 11(9):597–610

Krützfeldt J, Stoffel M (2006) MicroRNAs: a new class of regulatory genes affecting metabolism. Cell Metab 4(1):9–12

Lee EJ et al (2008) Systematic evaluation of microRNA processing patterns in tissues, cell lines, and tumors. RNA 14(1):35–42

Lee RC, Feinbaum RL, Ambros V (1993) The C. elegans heterochronic gene lin-4 encodes small RNAs with antisense complementarity to lin-14. Cell 75(5):843–854

Lee YS, Dutta A (2009) MicroRNAs in cancer. Annu Rev Pathol 4:199–227

Li Z et al (2011) Small RNA-mediated regulation of iPS cell generation. EMBO J 30(5):823–834

Lin N et al (2010) Profiling highly conserved microrna expression in recombinant IgG-producing and parental chinese hamster ovary cells. Biotechnol Prog. Available at: http://www.ncbi.nlm.nih.gov/pubmed/21692195. Accessed August 15, 2011

Lovis P, Gattesco S, Regazzi R (2008) Regulation of the expression of components of the exocytotic machinery of insulin-secreting cells by microRNAs. Biol Chem 389(3):305–312

Lu J (2005) MicroRNA expression profiles classify human cancers. Nature 435:834–838

Majors BS et al (2008) Enhancement of transient gene expression and culture viability using Chinese hamster ovary cells overexpressing Bcl-x(L). Biotechnol Bioeng 101(3):567–578

Majors BS et al (2009) Mcl-1 overexpression leads to higher viabilities and increased production of humanized monoclonal antibody in Chinese hamster ovary cells. Biotechnol Prog 25(4):1161–1168

Manni I et al (2009) The microRNA miR-92 increases proliferation of myeloid cells and by targeting p63 modulates the abundance of its isoforms. FASEB J 23(11):3957–3966

Medina PP, Nolde M, Slack FJ (2010) OncomiR addiction in an in vivo model of microRNA-21-induced pre-B-cell lymphoma. Nature 467(7311):86–90

Müller D, Katinger H, Grillari J (2008) MicroRNAs as targets for engineering of CHO cell factories. Trends Biotechnol 26(7):359–365

Nonne N et al (2010) Tandem affinity purification of miRNA target mRNAs (TAP-Tar). Nucleic Acids Res 38(4):e20

Nottrott S, Simard MJ, Richter JD (2006) Human let-7a miRNA blocks protein production on actively translating polyribosomes. Nat Struct Mol Biol 13(12):1108–1114

Olive V et al (2009) miR-19 is a key oncogenic component of mir-17–92. Genes Dev 23(24):2839–2849

Pickering MT, Stadler BM, Kowalik TF (2009) miR-17 and miR-20a temper an E2F1-induced G1 checkpoint to regulate cell cycle progression. Oncogene 28(1):140–145

Reddy SDN et al (2008) MicroRNA-7, a homeobox D10 target, inhibits p21-activated kinase 1 and regulates its functions. Cancer Res 68(20):8195–8200

Roobol A et al (2009) Biochemical insights into the mechanisms central to the response of mammalian cells to cold stress and subsequent rewarming. FEBS J 276(1):286–302

Suzuki HI et al (2009) Modulation of microRNA processing by p53. Nature 460(7254):529–533

Trummer E et al (2006a) Process parameter shifting: Part I. Effect of DOT, pH, and temperature on the performance of Epo-Fc expressing CHO cells cultivated in controlled batch bioreactors. Biotechnol Bioeng 94(6):1033–1044

Trummer E et al (2006b) Process parameter shifting: Part II. Biphasic cultivation-A tool for enhancing the volumetric productivity of batch processes using Epo-Fc expressing CHO cells. Biotechnol Bioeng 94(6):1045–1052

Vecchione A, Croce CM (2010) Apoptomirs: small molecules have gained the license to kill. Endocr-Relat Cancer 17(1):F37–50

Wightman B, Ha I, Ruvkun G (1993) Posttranscriptional regulation of the heterochronic gene lin-14 by lin-4 mediates temporal pattern formation in C. elegans. Cell 75(5):855–862

Wu C-I, Shen Y, Tang T (2009) Evolution under canalization and the dual roles of microRNAs—A hypothesis. Genome Res 19(5):734–743

Xu X et al (2011) The genomic sequence of the Chinese hamster ovary (CHO)-K1 cell line. Nat Biotechnol 29(8):735–741

Yee JC, de Leon Gatti M et al (2008) Genomic and proteomic exploration of CHO and hybridoma cells under sodium butyrate treatment. Biotechnol Bioeng 99(5):1186–1204

Yee JC, Wlaschin KF et al (2008) Quality assessment of cross-species hybridization of CHO transcriptome on a mouse DNA oligo microarray. Biotechnol Bioeng 101(6):1359–1365

Yee JC, Gerdtzen ZP, Hu W-S (2009) Comparative transcriptome analysis to unveil genes affecting recombinant protein productivity in mammalian cells. Biotechnol Bioeng 102(1):246–263

Yu X et al (2012) Induction of cell proliferation and survival genes by estradiol-repressed microRNAs in breast cancer cells. BMC Cancer 12:29

Zhu H et al (2011) The Lin28/let-7 axis regulates glucose metabolism. Cell 147(1):81–94

Zhu S et al (2008) MicroRNA-21 targets tumor suppressor genes in invasion and metastasis. Cell Res 18(3):350–359

Chapter 5
MicroRNAs as Engineering Targets: Pathway Manipulation to Impact Bioprocess Phenotypes

Aliaksandr Druz, Michael Betenbaugh and Joseph Shiloach

Abstract Chinese hamster ovary (CHO) cells are the primary mammalian culture system used for recombinant protein production; therefore there are continuous research and development efforts to improve cell production capabilities by both genetic modification and process optimization strategies. The genetic modifications are used to increase specific growth rate, to reduce apoptosis and to improve nutrients utilization. Since altering the expression of a single gene or even a single pathway may not be sufficient to produce desirable phenotypes, regulation of global gene expression may be a better approach for pathway engineering in CHO cells. miRNA(s) were found to be global regulators of gene expression with the ability to simultaneously alter multiple cellular pathways such as cell growth, apoptosis, stress resistance, metabolism and protein secretion. Therefore, modifications of miRNA expression profiles may facilitate the design of high-producing CHO cells. Recent advances in transfection techniques allow the insertion of miRNA mimics or inhibitors into CHO cells at specific stages of the bioprocess. Unlike traditional engineering approaches, manipulation of miRNA expression profiles does not burden the translational machinery of the cell and therefore, cellular metabolic resources are allocated to recombinant protein production. In this chapter we highlight the industrially-relevant pathways, report on miRNA involvement in their regulation, discuss how these miRNAs can be used to improve performance of CHO cells for industrial applications and propose specific miRNA candidates for CHO cell engineering.

Keywords Apoptosis · Mammalian cells · miRNA · Protein expression

---

J. Shiloach (✉)
9000 Rockville Pike Bldg 14A Rm 173, Bethesda, MD 20892, USA
e-mail: yossi@nih.gov

A. Druz · J. Shiloach
Biotechnology Core Laboratory NIDDK,
NIH Bldg 14A, Bethesda, MD 20892, USA

A. Druz · M. Betenbaugh
Department of Chemical and Biomolecular Engineering, Johns Hopkins University,
Baltimore, MD 21218, USA

N. Barron (ed.), *MicroRNAs as Tools in Biopharmaceutical Production*,
DOI 10.1007/978-94-007-5128-6_5,

5.1 Introduction

5.1.1 Potential Role of MicroRNAs in Improving CHO Cells Suitable for Large-Scale Production of Biologicals

Chinese hamster ovary cells are the primary mammalian culture system used for production of biopharmaceuticals. This is attributed to their robustness in a bioreactor and their ability to produce properly folded recombinant proteins with human-like posttranslational-modifications which is important for maintaining their biological activity (Barron et al. 2011b; Druz et al. 2011; Kramer et al. 2010; Muller et al. 2008).

Mammalian-based systems for biological production are required to produce high quantities of product and therefore need to meet process-related requirements that place high demands on CHO cells to generate suitable product titers and quality (Dinnis and James 2005). Improving and predicting the performance of CHO cells in bioreactors is important for enhancement of cell specific productivity (Clarke et al. 2011).

Limited growth capacities and low tolerance to different environmental conditions in bioreactors (nutrient and growth factor depletion, shear and oxidative stresses, metabolite accumulation, pH, osmolality and hypoxia) affect cell productivity. It is possible that these conditions cause apoptosis, reducing the yield and the quality of the produced proteins (Lim et al. 2010). Generation of stress-resistant CHO cell lines suitable for efficient production of various biologicals is therefore important for the biopharmaceutical industry. Additional desired characteristics of high-producing CHO cells are rapid growth, long term genomic stability and protein secretion capacity.

The discovery of microRNAs (miRNAs) offers an opportunity to improve the performance of CHO cells in an industrial-scale bioprocess. It is predicted that nearly half of the proteins may be affected by miRNAs (Lewis et al. 2005) which are more abundant than transcription factors. miRNAs have been shown to be global regulators of gene expression affecting almost all essential cellular processes and functions. So far, it has been found that miRNAs are involved in cell development, differentiation, metabolism and proliferation. miRNA expression profiles can also be explored as biomarkers for desired properties such as resistance to various stresses, growth and clonal stability and/or engineering targets (Barron et al. 2011a; Druz et al. 2011).

The potential of miRNAs utilization for mammalian cell engineering was highlighted by analyzing miR expression profiles under various physiological conditions (Druz et al. 2011; Gammell 2007). Each miRNA has the potential to post-trascriptionally affect more than 100 targets and concurrently influence several interlinked pathways or multiple points of the same pathway in the cell, which may facilitate creating industrially-relevant phenotypes (Barron et al. 2011a). One of the key advantages for utilizing miRNAs rather than regulatory proteins such as transcriptional factors or kinases is the fact that they do not burden the translational machinery and therefore reduce the metabolic load on the host cells. As a result, cellular metabolic resources will be better allocated to the production of recombinant protein, leading to increase of cell specific productivity. In addition, the cell can respond faster to miRNA modulations, either naturally or by introduction of exogenous mimics or inhibitors since there is no need for a translation step (Barron et al. 2011a; Muller et al. 2008).

Novel large scale transfection techniques allow relatively easy insertion of miRNA mimics or inhibitors into CHO cells at some stage of the bioprocess (Barron et al. 2011a). Also, some efficient miRNA expression systems which contain drug-inducible or heat-shock-inducible promoters that control the timing and expression levels of chosen miRNAs have been recently reported (Weber and Fussenegger 2007; Yang and Paschen 2008). Therefore, the investigation of miR expression profiles in CHO cells under different physiological conditions will facilitate detection of specific miRNAs that can be manipulated to obtain desired properties in CHO cells. Alteration of those miRNA expression profiles will introduce another dimension to genetic engineering of CHO cell to generate more robust CHO cell lines better suitable to production of biologics (Druz et al. 2011).

5.1.2 Current Strategies for CHO Cell Engineering

There is currently an effort to improve CHO cells' performance and productivity through better understanding of the physiology of the cells and their response to the physical and chemical environment in the bioreactor.

Current process optimization techniques focus on increasing integrated viable cell density (IVCD) in the bioreactor by enhancing the specific growth rate, reducing apoptosis, redirecting metabolic pathways to lower accumulation of toxic byproducts, and increasing nutrient utilization efficiency (Druz et al. 2011; Muller et al. 2008; Chen et al. 2001; Elias et al. 2003; Irani et al. 2002). Most cell line improvement efforts are based on media manipulation or gene engineering approaches by random integration in the cell. However, newly developed methods such as targeted homologous integration, RNA interference techniques and zinc-finger nuclease deletions should be evaluated as approaches for cell improvements (Kramer et al. 2010).

One of the main obstacles that researchers face while attempting to alter the CHO cells performance is the complexity of the intended phenotype such as rapid cell growth and resistance to various stresses in the bioreactor. Altering the expression of a single gene or protein or even a single pathway may not be sufficient to produce the desirable phenotype (Barron et al. 2011b). Therefore in some instances, cells have been "double engineered" to reach higher growth rates and increase apoptosis resistance (Fussenegger and Bailey 1998; Ifandi and Al-Rubeai 2005). It should be noted however that over-engineering of the cells may affect growth due to high metabolic load on the cells (Yallop and Svendsen 2001).

A comprehensive understanding of cellular pathway control and interaction is required for developing strategies to globally affect gene expression profiles for the purpose of engineering CHO cells with improved production properties (Dinnis and James 2005). Some success has been shown with the modification of the global gene expression levels through over-expression of transcription factors and artificial

Zn-finger proteins (ZFP). Over-expression of ATF4 and Xbp1 transcription factors and the artificial ZFP-TF derived from a synthetic library increased recombinant protein production capacity of CHO cells (Kwon et al. 2006; Ohya et al. 2008; Tigges and Fussenegger 2006). Dinnis and James (2005) suggested an approach based on imitating the molecular events of differentiation of B-lymphocytes into high producing plasma cells. This inverse engineering strategy enabled the synchronized expansion and maintenance of high levels of metabolic and secretory cellular machinery to increase cell-specific production rate in CHO cells.

Below we describe the industrially-relevant phenotypes and pathways with the existing strategies to improve performance of CHO cells, together with reports on miRNA involvement in the control of respective pathways. We concentrate on the possible manipulation of selected microRNAs in CHO cells to globally affect gene regulation and improve cell performance.

5.2 Applications of MicroRNA for Pathways Engineering in CHO Cells

5.2.1 Engineering of Growth and Growth Arrest

Volumetric production of mammalian cell culture is a function of cell specific productivity and IVC. Higher growth rates increase the rate of biomass accumulation and the number of viable cells, but suppressing cell growth can increase cell specific productivity by redirecting the metabolic energy from growth to recombinant protein production and secretion (Muller et al. 2008; Dinnis and James 2005). This conflicting relationship is managed by a "biphasic" cell culture strategy where in the first phase the cell growth is not limited, and in the second phase the cells growth is arrested to form high-producing cells. The challenge in this approach is to find a way to arrest cell division without promoting cell death or interfering with recombinant protein production (Dinnis and James 2005).

Techniques to induce growth arrest include manipulation of growth parameters (temperature, pH, hyperosmotic pressure), alteration of extracellular environment by adding specific metabolites or DNA synthesis inhibitors, and addition of cell-cycle regulators such as nucleotides or nucleosides (Altamirano et al. 2001; Bi et al. 2004; Carvalhal et al. 2003; Kim and Lee 2002). Each of these methods may affect multiple cellular processes simultaneously. For example, reduction of culture temperature was reported to suppress cell growth in a biphasic process which increased product titer in CHO cells; at the same time, lowered temperature induces changes in gene expression, protein phosphorylation, nucleotide pools, and consequently a reduction in cell metabolism (Kaufmann et al. 1999; Yoon et al. 2003). In addition, the effects of reduced temperature may be cell or product-specific and cannot be generalized (Dinnis and James 2005; Yoon et al. 2003).

Genetic control of cell cycle progression is another strategy to induce growth arrest. This can be done by activation of intrinsic cell cycle modulators including genes encoding the p21^{CIP1} and p27^{KIP1} cyclin E-dependant kinase cdk2 inhibitors which cause arrest of the cell cycle at G1 phase (Fussenegger and Bailey 1998). This strategy of cell cycle arrest may also have its own limitations since it is difficult to engineer higher cell growth rates and growth arrest simultaneously. Besides, the improvement of cell specific productivity resulting from this approach may also be cell and product-specific (Dinnis and James 2005).

A microRNA-based strategy is a possible alternative to the existing methods of manipulating cell growth and arrest. There are reports on miRNA involvement in cell proliferation and growth by targeting essential genes of cell cycle regulation. The function of several members of miRNA 17–92 polycistronic cluster was investigated. Inhibition of miR-18a with antagomirs decreased cell proliferation (Scherr et al. 2007) and miR-17–5p and miR-20a negatively regulated the expression of E2F1 protein (which, when up-regulated promotes cell cycle progression and overrides the growth arrest), (O'Donnell et al. 2005). Another study showed that the inhibition of miR-21 and miR-24 resulted in a significant increase in mammalian cell growth (Cheng et al. 2005). The growth-inhibitory properties of miR-21 and miR-24 were also confirmed by their up-regulation during the stationary phase and after a temperature shift when compared to their levels during the exponential growth phase (Gammel 2007). It was suggested that miR-24 inhibits the progression thorough G1 phase and transition to S phase of the cell cycle by targeting E2F2 and MYC, and may also regulate DNA repair (Lal et al. 2009).

Several miRNAs have been reported to act as regulators of cell growth cycle. (1) miR-125b was reported to support cell growth in liver cells, and its inhibition in human cancer cell lines caused decreased growth (Gammell 2007; Lee et al. 2005); (2) over expression of Let-7 suppressed growth in colon cancer and lung adenocarcinoma by targeting MYC and RAS gene expression, thus affecting the pathways controlled by these oncogenes (Akao et al. 2006; Johnson et al. 2005); (3) expression of miR-133 was found to be correlated with a higher skeletal muscle growth rate by repression of serum response factor (SRF) in cultured myoblasts (Chen et al. 2006); (4) the miR-34 family (miR-34 a, b, c) was found to target growth and growth-arrest related genes including CDK4,and CDK6, c-MYC, CREB, Notch1, E2F3 (Hermeking 2010).

Other miRNAs affect proliferation of mammalian cell lines: (1) miR-143 and miR-145 negatively affect cell proliferation by regulating genes of the mitogen-activated protein kinase (MAPK) family, MAPK7 and ERK5 (Esau et al. 2004); (2) miR-221 and miR-222 inhibit proliferation in hematopoietic progenitor cells lines by targeting the Kit gene, which encodes receptor protein with tyrosine-kinase activity (Felli et al. 2005); (3) miR-372 and miR-373 control proliferation by direct interference with cell cycle regulation, regardless of increased levels of the inhibitor p21^{CIPI} and inhibition of CDK by p53; (4) over expression of miR-31 exhibits anti-proliferative effects in multiple cancer types by associating with the p53-mediated growth inhibitory pathway, and may trigger caspase activation and apoptosis by suppressing E2F2 gene (Creighton et al. 2010); (5) over expression of miR-7 inhibited growth by targeting epidermal growth factor receptor (EGFR) and Akt pathway in glioblastomas. Exogenous increase of miR-7 levels blocked CHO-K1 proliferation and increased cell productivity, suggesting its possible role in CHO cells engineering via decoupling cell growth from growth arrest to enhance recombinant protein production (Barron et al. 2011a; Kefas et al. 2008).

The use of miRNAs for engineering CHO cells better suitable for growth and production needs to be evaluated carefully since their effects may be tissue or cell-specific (Gammel 2007). miR-21 and miR-24 have already been reported in CHO cells and seem to have growth-inhibitory properties (Gammel 2007). Recent work on sequencing the CHO microRNA transcriptome will serve as a valuable resource for identification of specific miRNAs involved in control of cell proliferation and cell cycle. Next-generation sequencing of miRNAs across different industrially-relevant conditions such as temperature shift and sodium butyrate treatment, helped to identify and annotate 387 mature miRNAs conserved in CHO including the members of miR 17–92 polycistronic cluster, miR-221/222 cluster, let-7 family, miR-34 family, miR-7, miR-125b, miR-143, and miR-31 (Hackl et al. 2011; Johnson et al. 2011). The sequencing of cDNA of miR 17–92 cluster validated targets confirmed the conservation of respective miRNA binding sites in 19 genes in CHO. The latter serves as a confirmation of the conserved biological role of this cluster in CHO cells (Hackl et al. 2011). Further studies of biological functionality of the reported CHO-specific miRNAs as compared to their role in other mammalian cells will facilitate identification of key miRNAs for growth and cell cycle engineering in CHO cell factories.

5.2.2 Engineering of Apoptosis-Resistant Cell Lines to Increase IVC

Apoptosis or programmed cell death (PCD) is a necessary physiological function in multicellular organisms that presents difficulty for maintaining high viable cell densities in mammalian bioprocess applications (Muller et al. 2008). Different stress conditions in bioreactors such as nutrient limitation, byproduct accumulation, shear and oxidative stresses, pH, osmolality and hypoxia can trigger apoptosis during CHO cell cultures. Onset of apoptosis results in a lowered IVC which affects product yield and properties (Druz et al. 2011; Gammel 2007). As a result, apoptosis prevention is one of the most investigated techniques in CHO cells engineering.

There are two main approaches to inhibit or slow the apoptotic cascade activation. The first involves the manipulation of the outer cellular environment by media supplementation with growth factors, limiting nutrients, and hydrolysates (Majors et al. 2007; Zanghi et al. 1999). The second employs genetic engineering methods to reallocate the tightly regulated balance of pro- and anti-apoptotic factors in favor of the anti-apoptotic proteins (Chiang and Sisk 2005; Lim et al. 2006; Wong et al. 2006). Anti-apoptotic proteins such as Bcl-x_L, Bcl-2 and Mcl-1 are known to protect the cells from apoptosis by maintaining the integrity of mitochondrial membrane. Exogenous expression of these proteins increases cell density and viability and protects the culture from apoptosis induction by different stimuli in mammalian cells (Fassnacht et al. 1999; Fussenegger et al. 2000; Mastrangelo et al. 2000;

Sauerwald et al. 2006; Tey et al. 2000). Also over-expression of viral homologs of bcl-2 such as E1B-19K was shown to reduce apoptosis in mammalian cell cultures (Figueroa et al. 2007; Mercille and Massie 1999; Sauerwald et al. 2002).

RNAi silencing techniques against pro-apoptotic genes have been successfully applied. Stable inhibition of Bax and Bak with shRNA vectors, and silencing of the apoptosis linked gene Alg-2 and the transcriptional factor Requiem in CHO cells delayed apoptosis, increased cell viability and improved interferon-γ production (Lim et al. 2006; Wong et al. 2006). Also, the simultaneous down-regulation of caspase-3 and caspase-7 expression improved cell viability and thrombopoietin production in CHO cells after sodium butyrate treatment (Sung et al. 2007).

miRNA involvement in apoptosis regulation was initially studied in peripheral blood cells of people diagnosed with chronic lymphocytic leukemia (CLL) where deletion of miR-15 and miR-16 genes was reported in the majority of patients (Calin et al. 2002). Later studies revealed that these miRNAs were promoting apoptosis in malignant B cells by targeting Bcl-2 expression at the post-transcriptional level (Cimmino et al. 2005). miR-21 (in addition to its involvement in cell growth regulation) was found to be up-regulated in several human cancers and was characterized as an oncogenic miR. Silencing of miR-21 expression in glioblastoma cells led to increased apoptosis by activation of caspase 3 and 7 (Chan et al. 2005; Meng et al. 2006; Si et al. 2007).

Cheng et al. have identified several miRNAs involved in apoptosis regulation by using large-scale antisense miRs inhibition. The inhibition of miR-1d, 7, 148, 204, 210, 216 and 296 in HeLa cells increased apoptosis by activation of caspase 3, while the inhibition of miR-214 had the opposite effect (Cheng et al. 2005)

Li et al. have investigated the pro-apoptotic role of miR-204 in human trabecular meshwork (HTM) cells. This miR directly targeted several anti-apoptotic genes and increased cell susceptibility to apoptosis (Li et al. 2011).

Other miRNAs have a role in the regulation of apoptosis: (1) miR-218 was found to be involved in NF-kappaB response and apoptosis induction by targeting expression of the ECOP gene (Gao et al. 2010); (2) miR-1 and miR-133 produced opposing effects on apoptosis induced by oxidative stress in rat cardiomyocytes (Xu et al. 2007). miR-1 had a pro-apoptotic function in response to oxidative stress by targeting heat shock proteins HSP60 and HSP70 and miR-133 seemed to have an anti-apoptotic role by repressing caspase 9 gene expression; (3) miR-34 family members were found to function as potent mediators of the p53-induced apoptotic pathway by targeting anti-apoptotic genes including Bcl-2, and were also found to participate in a positive feedback loop of p53 activation via increased acetylation by targeting SIRT1 deacetylase (Hermeking 2010); (4) miR-30 was shown to affect the levels of the Ubc9 and ITGB3 genes in breast tumor-initiating cells which restricted their self-renewing capacity and targeted them for apoptosis (Yu et al. 2010); (5) miR-10a was shown to participate in the TRAIL-induced apoptosis pathway leading to caspase 3 activation in human lung carcinomas (Ovcharenko et al. 2007); (6) the members of let-7 miRs family, let-7c and let-7g, were shown to target Bcl-x_L directly and Mcl-1 indirectly which led to caspase-3/7 activation and apoptosis induction in hepatocytes (Shimizu et al. 2010).

Druz et al. recently showed the up-regulation of the large miR 297–669 cluster during apoptotic conditions induced by nutrient depletion in CHO cells. One member of this cluster, miR-466h was shown to alter the expression of five anti-apopototic genes from different apoptosis-initiating pathways (bcl2l2, dad1, birc6, stat5a and smo). Antisense knockdown of miR-466h delayed apoptosis onset in nutrient-depleted conditions by decreasing caspase activation and increasing the cell viability (Druz et al. 2011). One other member of the miR 297–669 cluster, miR-669c, had been previously associated with impairments in glutathione metabolism which activated the apoptosis cascade (Lanceta et al. 2010; Maes et al. 2008).

Application of apoptosis regulation-related miRNAs in the engineering of CHO cells is possible; it should provide researchers with another tool for apoptosis inhibition. The recent sequencing of the CHO cell genome and miRNA transcriptome will provide information on gene down-regulation in CHO cells by CHO-specific apoptosis-related miRNAs (Hackl et al. 2011; Johnson et al. 2011; Xu et al. 2011). The sequencing of CHO-specific miRNAs has revealed a conserved sequence of the apoptosis-regulating miRNAs such as the miR-15/16 cluster, let-7 family, miR-214, miR-218, miR-1, and miR-10a. These miRNAs and the miR 297–669 cluster are possible targets for apoptosis pathway engineering. Due to the complexity of the PCD and the diversity of the apoptotic stimuli, it may be useful to investigate the combined effects of several miRNAs affecting genes from different stages of the apoptosis cascade, and the effects of miRNAs which seem to be involved in global regulation of the pathway. In addition, it might be worthwhile to consider the engineering of whole clusters of apoptosis-relevant miRNAs, since clustered miRNAs are known to be transcribed together as polycistronic transcripts to regulate the mRNA of genes with similar functions (Druz et al. 2011). More studies on CHO cell-specific miRNAs, their biological role and their effects on CHO cell-specific gene targets need to be conducted to suggest the most suitable miRNA candidates to be used in industrial scale bioprocesses.

5.2.3 *Role of MicroRNA in Engineering of Stress Response*

Stress conditions in bioreactors do not only induce apoptosis but can also affect product yield and properties (Muller et al. 2008). For example, high protein expression and limited secretion can lead to accumulation of aggregated and misfolded proteins, which can generate stress in the endoplasmic reticulum (ER). This stress was shown to be reduced by ER expansion via ectopic expression of transcriptional factor X-box-binding protein1, Xbp1 (Tigges and Fussenegger 2006), and over-expression of the protein disulfide isomerase (PDI) (Borth et al. 2005). Over-expression of ER chaperones, calreticulin and calnexin (induced by heat shock, amino acid deprivation and perturbations of Ca^{2+} levels) increased cell specific productivity of recombinant thrombopoietin in CHO cells (Chung et al. 2004; Heal and McGivan 1998).

Dissolved oxygen gradients may affect growth, metabolism, and the recombinant protein production. The gradients can be significant in bioreactors as a result of inefficient mixing that exposes the cells to different oxygen tensions ranging from absence of oxygen (hypoxia) to above-atmospheric concentrations (hyperoxia) (Dunster et al. 1997). An example of a genetic engineering approach to reduce oxidative stress is the over-expression of the antioxidant haptoglobin which increased tolerance to oxidative stress in CHO cells (Tseng et al. 2004). The effects of shear stress on cell growth, metabolite consumption, and protein production were investigated in recombinant CHO cell cultures producing tissue plasminogen activator (tPA) and human growth hormone (hGH) (Keane et al. 2003; Senger and Karim 2003) In both cases, shear stress had a negative effect on recombinant protein production. Elevated levels of ammonia were identified as another stress condition that lowers the expression of genes managing cell cycle and protein folding, up-regulates genes affecting energy metabolism and induces protein degradation (Chen and Harcum 2006; 2007). Cold stress was shown to have positive effects on recombinant protein production in CHO cells (Fox et al. 2005) and the exogenous over-expression of the cold-inducible RNA-binding protein (CIRP) increased the titer of recombinant protein (Tan et al. 2008).

Changes in expression profiles of several miRNAs have been observed in response to ER and hypoxic stress conditions: (1) miR-708 was shown to be induced during ER stress by the transcription factor CCAAT enhancer-binding homologous protein (CHOP), and may facilitate the enhancement of ER protein-folding capacity under the stress of accelerated protein synthesis (Behrman et al. 2011); (2) miR-204 supports ER and oxidative stress induction in human trabecular meshwork cells. This miRNA inhibited two genes involved in the elimination of damaged and misfolded proteins (SERP1/RAMP4 and M6PR) and facilitated the increase of carbonylated proteins (Li et al. 2011); (3) The miR-221/222 cluster was down-regulated during ER stress in human hepatocellular carcinoma cells. The ectopic introduction of miR-221/222 mimics increased ER-stress induced apoptosis which was associated with $p27^{Kip1}$ and MEK/ERK-directed cell cycle regulation (Dai et al. 2010); (4) miR-15a, miR-16, and miRs-20a were down-regulated at hypoxic conditions in human carcinomas (Hua et al. 2006); (5) miR-26, miR-107, and miR-210 were up regulated in neoplastic cells in response to low oxygen. These miRNAs are likely to decrease the pro-apoptotic signaling in a hypoxic environment (Kulshreshtha et al. 2007). miR-210 was also found to be progressively up-regulated in endothelial cells in hypoxic conditions and inhibited receptor tyrosine-kinase ligand Ephrin-A3 which is critical in vascular development (Fasanaro et al. 2008); (6) The up-regulation of the miR-34 family, while being part of the p53 network, can be also implicated as a stress response to DNA damage, hyperactive cytokine signaling, and hypoxia (He et al. 2007); (7) The miR 17–92 cluster was shown to target hypoxia-inducible factor alpha (Hif-1α), a transcriptional factor known to regulate cellular response to hypoxia. The latter has an important role in various biological processes such as glucose metabolism, pH regulation and angiogenesis (Taguchi et al. 2008); (8) miR-31 was shown to activate Hif-1α via the inhibition of factor-inhibiting hypoxia -inducible factor (HIF) (Liu et al. 2010).

Several miRNAs were found to be associated with oxidative stress: (1) the bicistronic transcript miR-144/451 is involved in oxidative stress in erythroid cells. miR-144 was shown to modulate the oxidative stress response in K562 and primary erythroid progenitor cells by directly affecting NRF2 gene expression which induces the expression of several antioxidant enzymes (Sangokoya et al. 2010). miR-451 protects the erythrocytes against oxidative stress and reverses differentiation defect of erythroid cells by inhibiting the intracellular regulator of cytokine signaling, 14–3-3ξ gene (Patrick et al. 2010). (2) miR-34a and miR-93 were involved in the loss of oxidative stress defense, and repress expression of genes associated with oxidative stress regulation and defense mechanism such as Sp1, Sirt1, Mgst1, and Nrf2 (Li et al. 2011).

Some miRNAs have been associated with other stress inducers in a bioreactor such as osmostic pressure, shear stresses, and nutrient depletion /gradients: (1) miR-200b and miR-717 are down-regulated by isotonic and hypertonic treatments in renal medullary epithelial cells. However, when up-regulated, these miRs inhibit the activity of transcriptional factor called osmotic response binding protein, OREBP, a major cellular osmoregulator in kidney cells and T-lymphocytes (Huang et al. 2010); (2) miR-7b was over-expressed in hyperosmolar conditions to down-regulate the protein levels of Fos. This reduces the activity of transcription factor activator protein 1 (AP1), a regulator of cellular processes, which is formed by the dimerization of Fos and Jun proteins (Lee et al. 2006); (3) miR-21 and miR-19a were induced by shear stress in endothelial cells (Qin et al. 2010; Weber et al. 2010); (4) Members of miR 297–669 cluster were up-regulated in response to nutrient depletion in CHO cells (Druz et al. 2011).

The utilization of stress-related miRNAs should be considered for bioprocess-relevant pathway engineering to eliminate negative effects on recombinant protein production. The homologs for some stress-related miRNAs (miR-7b, 93, 107, 144, 200b, 210, 708) were reported in CHO cells via next-generation sequencing (Hackl et al. 2011) and their activity can be examined in CHO cells.

5.2.4 Metabolic Disorders and Their Prevention

Accumulation of lactate and ammonia can significantly affect cell growth and production capability (Ozturk et al. 1992). Lactate is a major by-product of glucose and glutamine metabolism. When glucose is converted to lactate, glutamine metabolism supports the TCA cycle by its conversion to glutamate and later to α-ketoglutarate, which generates excessive amounts of ammonia (Muller et al. 2008). Ammonia accumulation was shown to affect cell growth and the properties of secreted protein via altered glycosylation pattern in CHO cells (Yang and Butler 2000).

The main approaches to optimize metabolic performance of mammalian cells include optimization of feeding strategies and genetic engineering of specific metabolic pathways (Irani et al. 2002). Over-expression of cytosolic pyruvate carboxylase (PC) enzyme in BHK cell cultures reduced lactate accumulation, and improved glucose and glutamine metabolism and recombinant protein production in a broad range of glucose concentrations in culture media (Elias et al. 2003; Irani et al. 2002). Knock-down of lactate dehydrogenase A enzyme (LDH-A), which catalyzes conversion of pyruvate to lactate, reduced glycolytic flux and the specific lactate production rate in CHO cells (Kim and Lee 2007). However, despite the reduced lactate accumulation and increased rate of oxidative phosphorylation, CHO cells became more susceptible to oxidative stress after knock-down of the LDH-A (Jeong et al. 2004). Korke et al. showed that a lowered ratio of glucose consumption to lactate production in hybridoma cells culture was the result of global changes in gene expression at the transcription level (Korke et al. 2004) and may be in part due to regulation of genes by miRNAs (Muller et al. 2008).

Another tool to investigate metabolic phenotypes is a metabolomics approach coupled with ^{13}C-flux analysis of cellular metabolism and its regulation in mammalian cells. This approach offers the quantification of internal metabolic fluxes, providing researchers with comprehensive information on cellular energetics (Zamboni and Sauer 2009).

Several miRNAs have been related to metabolic control. (1) miR-375 was shown to regulate glucose homeostasis and glucose-mediated insulin secretion in pancreatic endocrine cell lines by targeting the myotrophin (Mtpn) gene (Poy et al. 2007). miR-375 was also shown to down-regulate 3′-phosphoinositide-dependent protein kinase-1 (PDK-1), resulting in decrease of insulin gene expression in primary rat islets (El Ouaamari et al. 2008). Over-expression of miR-375 decreased glucose-induced insulin secretion with no effects on glucose-stimulated ATP production or intracellular Ca^{2+} levels (Gauthier and Wollheim 2006); (2) miR-124a, miR-107, miR-30d were up-regulated, and miR-296, miR-484, miR-690 were down-regulated at high glucose conditions in pancreatic beta cells. Over-expression of miR-30d increased insulin gene expression indirectly, but had no effects on insulin secretion (Tang et al. 2009); (3) miR-122 was shown to be involved in regulation of cholesterol and lipid homeostasis in mice (Esau et al. 2006; Krutzfeldt and Stoffel 2006); (4) miR-29b involvement in amino acid metabolism was shown in HEK293 cells. This miRNA controlled the branched amino acid (BCAA) metabolism by targeting the branched-chain α-ketoacid dehydrogenase (BCKD) enzyme, known to catalyze the irreversible step in BCAA catabolism (Mersey et al. 2005); (5) Gao and colleagues showed the effect of miR-23a/b in human B lymphoma cells on the regulation of glutamine metabolism by targeting mitochondrial glutaminase (GLS) expression. It was also shown that miR-23a/b is subject to c-MYC regulation (Gao et al. 2009); (6) miR-378*, was shown to induce the metabolic shift in breast cancer cells by targeting estrogen related receptor (ERRγ), and GA-binding protein α (GABPA), two key regulators of energy metabolism. Over-expression of miR-378* reduced the activity of the TCA cycle, rendering the cells less dependent on oxidative phosphorylation and causing increased lactate production (Eichner et al. 2010).

Some metabolism-related miRNAs discussed above (miR-23a/b, -29b, -30d, -107, -122, -296, -484, -378*) were already reported in CHO cells (Hackl et al. 2011; Johnson et al. 2011). miR-dependent regulation of metabolic pathways is relatively new; therefore, comprehensive analysis of miR-mediated control of metabolic enzymes and fluxes and their effects on metabolic phenotypes, coupled with the investigation of CHO-specific miRs and gene targets, needs to be conducted.

5.2.5 Protein Secretion Pathway Engineering

Recombinant protein production in mammalian cells is achieved by controlling recombinant gene expression and optimizing protein secretion. Improving gene expression has received considerable attention, and approaches such as codon optimization and promoter and enhancer sequence design have been utilized (de Boer et al. 2004; Kim and Lee 1997; Masuda et al. 2000). However, limited attention has been given to protein secretion engineering (Muller et al. 2008).

One way to improve the secretion of the recombinant protein is over-expression of the spliced form of X-box binding protein1 (Xbp-1S), a transcriptional activator for a number of genes linked to protein secretion and endoplasmic reticulum (ER) biosynthesis. Over-expression of Xbp-1S in mammalian cells increased recombinant protein secretion in cases when protein accumulation exceeded the secretory capacity of the cells (Ku et al. 2008).

Kantardjieff et al. showed direct correlation between enhanced protein production and overall increase of ER and Golgi-related protein secretion system in CHO and NS0 cells at low temperature and sodium butyrate treatment (Kantardjieff et al. 2010). Therefore, the genes that are closely engaged in the secretory pathway are worth considering as potential targets for secretion pathway engineering. Suitable candidates include coat protein complexes I and II, which recruit the cargo and direct its traffic through the early secretory pathway between ER and Golgi (Muller et al. 2008; Barlowe 2000). Over-expression of Munc 18b, the regulator of the fusion of secretory vesicles with the plasma membrane, increased heterologous protein production in several mammalian cell lines (Peng 2010). In addition, the ectopic expression of the synaptosome-associated protein (SNAP-23) and vesicle-associated membrane protein 8 increased mammalian cell productivity (Peng et al. 2011).

Few miRNAs have been associated with controlling the secretory pathway in mammalian cells: (1) miR-124a and miR-96 affected the genes involved in insulin secretion in pancreatic β-cells (Lovis et al. 2008). miR-124a increased insulin exocytosis at low glucose concentrations by indirectly increasing the levels of SNARE protein SNAP25, GTPase Rab3A, and synapsin-1A and by directly targeting the GTPase Rab27A. miR-96 increased the levels of granuphilin protein, a potent inhibitor of insulin exocytosis, and decreased the expression of Rab27A-binding protein, Noc2 (Lovis et al. 2008); (2) miR-9 reduced glucose and potassium-dependent insulin exocytosis by targeting transcription factor Onecut2 (OC2) which represses Granuphilin/Slp4, that negatively affects the insulin secretion (Plaisance et al. 2006); (3) miR-375 has negative effects on the glucose-mediated insulin secretion (Gauthier and Wollheim 2006).

Although most of the knowledge about miRNA involvement in protein secretion comes from studies of pancreatic β-cells, there are some reports of miRNA connection to the senescence -associated secretory phenotype, characterized by growth arrest, increased gene expression and secretion of various proteins (Bhaumik et al. 2009). Bhamuik et al. showed up-regulation of miR-146a/b in senescent human fibroblasts. The ectopic expression of miR-146a/b was shown to target a component of IL-1 transduction pathway and reduced the secretion of IL-6 and IL-8, thereby participating in the negative feedback loop which suppresses the imbalance of the senescence -associated secretory phenotype activity (Bhaumik et al. 2009).

The engineering of protein secretory pathway via manipulation of miRNA levels is feasible for industrial scale applications in CHO cell cultures, especially considering that homologs for miR-9 and miR-146a/b have been reported in CHO cells (Hackl et al. 2011; Johnson et al. 2011). This research is in its early stages and more studies are needed to identify the effects of manipulating specific genes on the recombinant protein secretion rates, and to identify the CHO-specific miRNAs involved in targeting those genes.

5.3 Global Modification of MicroRNA Expression to Improve CHO Cells Bioprocesses

Global modification of miRNA expression levels is another strategy to improve CHO cell performance. The fact that mature miRNA expression levels in tumors are usually lower than their levels in healthy tissues (Lu et al. 2005), and over-expression of miRNAs frequently prevents transformation of cells to cancer-like phenotype, indicates their possible anti-oncogenic role (Imam et al. 2010; Lee et al. 2010). Therefore, the global down-regulation of miRNA expression may improve proliferative capacity of the cells and eliminate the stress response activation pathways (Barron et al. 2011a).

Modification of miRNA expression levels by genetic or epigenetic means, and by varying the levels of the miRNA biosynthesis enzymes and degradation factors, should be considered for CHO cell engineering (Barron et al. 2011a). An example of genetic regulation of miRNA profiles is the effect of the oncogenic transcription factor, c-Myc, which down-regulates miRNA expression levels by direct association with promoters of pri-microRNAs (Chang et al. 2008). Another study showed the opposite effect, where c-Myc activates the pro-oncogenic miR 17–92 cluster by direct binding to this cluster's promoter (O'Donnell et al. 2005). Other transcription factors, such as Hif-1alpha, NF-κB and p53 were also shown to affect expression of miRNA clusters by interaction with their promoters (Sun et al. 2010).

Epigenetic regulation of a single miRNA or an miRNA cluster can be initiated by DNA methylation and histone modifications; it was shown by the activation of epigenetically silenced miR-127 in human cancer cell lines (Saito et al. 2006). It is also possible to alter miRNA biogenesis by knocking down the miRNA processing enzymes Dicer and Drosha, or by over-expression of degradation factors such as XRN2 (Martello et al. 2010).

The global modification of miRNA expression needs to be carefully investigated to avoid potential harmful effects on cell homeostasis. For example, even though knockout of Dicer and Drosha can be implemented to globally down-regulate miRNA expression, it can negatively affect cell-cycle progression and mitosis in some cell lines (Martello et al. 2010). More research needs to be conducted to clarify the effects of global miRNA alterations on cellular functions in CHO cells.

Table 5.1 Involvement of miRs in industrially-relevant pathways

microRNA	Reported in CHO	Affected pathway	Effect type	Target(s)
miR-1	Yes	Apoptosis	Pro-apoptotic	HSP60, HSP70
miR-1d	No	Apoptosis	Anti-apoptotic	?
miR-7	Yes	Cell growth	Growth inhibitory	EGFR
		Apoptosis	Anti-apoptotic	?
miR-7b	Yes	Osmotic stress	Inhibition of stress response	Fos
let-7	Yes	Cell growth	Growth inhibitory	c-Myc, RAS
let-7c/g	Yes	Apoptosis	Pro-apoptotic	Bcl-x_L Mcl-1
miR-9	Yes	Protein secretion	Inhibits protein secretion	OC2
miR-10a	Yes	Apoptosis	Pro-apoptotic	?
miR-15a@ miR-16	Yes	Apoptosis	Pro-apoptotic	Bcl-2
		Hypoxic stress	Possible induction of apoptosis under hypoxia	?
miR-17-5p@ miR-20a	Yes	Cell growth	Inhibit/promote growth	E2F1
		Hypoxic stress	Possible induction of apoptosis under hypoxia	Hif1-alpha
miR-18a	Yes	Cell growth	Promotes growth	?
miR-19a	Yes	Shear stress	Anti-proliferative effect under laminar shear stress	Cyclin D1
miR-21	Yes	Cell growth	Growth inhibitory Anti-apoptotic	CDC25A
		Apoptosis	Decrease in apoptosis under laminar shear stress	PDCD4
		Shear stress		PTEN
miR-23a/b	Yes	Metabolism	Control of glutamine metabolism	GLS
miR-24	Yes	Cell growth	Growth inhibitory	E2F2, c-MYC
miR-26	Yes	Hypoxic stress	Decrease pro-apoptotic signals during hypoxia	?
miR-29b	Yes	Metabolism	Control of branched amino acids metabolism	BCKD
miR-30	No	Apoptosis	Pro-apoptotic	Ubc9, ITGB3
miR-30d	Yes	Metabolism	Regulation of glucose metabolism	?
miR-31	Yes	Cell growth	Growth inhibitory	E2F2
		Hypoxia	Possible resistance to apoptosis under hypoxia	FIH
miR-34a/b/c	Yes	Cell growth	Growth inhibitory	CDK4,6, c-MYC, CREB, Notch1, E2F3
		Apoptosis	Pro-apoptotic	Bcl-2, SIRT1
		Stress response	Possible induction of apoptosis under stresses	Sp1, Sirt1, Mgst1, Nrf2
miR-93	Yes	Oxidative stress	Loss of oxidative stress defense	Sp1, Sirt1, Mgst1, Nrf2
miR-96	No	Protein secretion	Inhibits protein secretion	Noc2
miR-107	Yes	Hypoxia	Decrease pro-apoptotic signals during hypoxia	?
		Metabolism	Regulation of glucose metabolism	?
miR-122	Yes	Metabolism	Regulation of cholesterol and lipid metabolism	?

Table 5.1 (continued)

microRNA	Reported in CHO	Affected pathway	Effect type	Target(s)
miR-124a	No	Metabolism	Regulation of glucose metabolism	?
			Promotes protein secretion	Rab27A
miR-125b	Yes	Cell growth	Promotes growth	?
miR-133	No	Cell growth	Promotes growth	SRF
miR-143	Yes	Cell growth	Growth inhibitory	ERK5
miR-146a/b	Yes	Protein secretion	Restrains protein secretion	NFκB, IRAK1, TRAF6
miR-148	Yes	Apoptosis	Anti-apoptotic	?
miR-144@ miR451	Yes	Oxidative stress	Protection against oxidative stress	Nrf2
				14-3-3ξ
miR-204	Yes	Apoptosis	Anti/pro-apoptotic	Bcl2l2, BIRC2, EZR, M6PR, SERP1
		ER and oxidative stress	Stress induction	M6PR
miR-200b	Yes	Osmotic stress	Inhibition of stress response	CREBP
miR-210	Yes	Apoptosis	Anti-apoptotic	?
		Hypoxia	Decrease pro-apoptotic signals during hypoxia	Ephrin-A3
miR-214	Yes	Apoptosis	Pro-apoptotic	?
miR-216	No	Apoptosis	Anti-apoptotic	?
miR-218	Yes	Apoptosis	Pro-apoptotic	ECOP
miR-221@ miR222	Yes	Cell growth	Growth inhibitory	Kit
		ER stress	Increased ER stress-induced apoptosis	?
miR-296	Yes	Apoptosis	Anti-apoptotic	?
		Metabolism	Regulation of glucose metabolism	?
miR-372@ miR-373	No	Cell growth	Promotes growth	LATS2
miR-375	No	Metabolism	Regulation of glucose metabolism	Mtpn, PDK-1
		Protein secretion	Inhibits protein secretion	?
miR-378*	Yes	Metabolism	Reduced activity of TCA cycle	ERRγ, GABPA
miR-466h	Yes	Apoptosis	Pro-apoptotic	Bcl2l2, Dad1, BIRC6, Stat5a, Smo
miR-484	Yes	Metabolism	Regulation of glucose metabolism	?
miR-690	No	Metabolism	Regulation of glucose metabolism	?
miR-708	Yes	ER stress	Improve protein folding	RHO
miR-717	No	Osmotic stress	Inhibition of stress response	OREBP

miRs effects on industrially-relevant pathways and their respective targets. (@) indicates combines miRs with the same effect on the pathway and same targets, (?) indicates unknown miR targets in relation to specific pathways

5.4 Conclusions

In this chapter we reviewed the potential role of miRNAs in CHO cell engineering for industrial applications based on their involvement in regulation of relevant pathways in other cellular systems. Table 5.1 summarizes the effects of the reviewed miRNAs on relevant cellular pathways and targeted genes in various systems. The recent work on CHO miRNA transciptome sequencing (Hackl et al. 2011; Johnson et al. 2011) allowed detection of more than 80 % of all miRNAs shown in Table 5.1. Only a few miRNAs have been studied in CHO cells so far: miR-7, miR-21/24 and miR-466h. The potential use of miR-7 and miR-466h has already been established by the manipulation of their expression levels in CHO (Druz et al. 2011; Barron et al. 2011b). Other promising individual miRNAs for CHO cell engineering include those involved in regulation of multiple pathways: miRs 15a and 16, miRs 17–5p and 20a, miRs 221 and 222, miR-34 family, miR-21, miR-31, miR-107, miR-124a, miR-204, miR-210, miR-296, and miR-375. The availability of the CHO cell genome sequence and the creation of CHO-specific genomic and proteomic tools will help to explore the activity of these miRNAs and to identify their targets. Another approach is to re-engineer (over-express or knock down) whole miRNA clusters already reported to be involved in relevant pathways such as miR 15a/16, miR 221/222, miR 144/451, miR 17–92, and miR-34b/c clusters. Also, clusters that incorporate the individual miRs shown in Table 5.1, such as miR 297–669, miR-296/298, miR 23–27 (include miR-466h, miR-296, and miR-23a respectively) need to be considered for the generation of robust CHO cell lines.

Acknowledgments Funding was provided by the intramural program of the National Institute of Diabetes and Digestive and Kidney Diseases, National Institutes of Health. The authors would like to thank Mrs. D. Livnat and Mrs. A Shiloach for critical review of the manuscript.

References

Akao Y, Nakagawa Y, Naoe T (2006) let-7 microRNA functions as a potential growth suppressor in human colon cancer cells. Biol Pharm Bull 29(5):903–906

Altamirano C, Cairo JJ, Godia F (2001) Decoupling cell growth and product formation in Chinese hamster ovary cells through metabolic control. Biotechnol Bioeng 76(4):351–360

Barlowe C (2000) Traffic COPs of the early secretory pathway. Traffic 1(5):371–377

Barron N et al (2011a) Engineering CHO cell growth and recombinant protein productivity by overexpression of miR-7. J Biotechnol 151(2):204–211

Barron N et al (2011b) MicroRNAs: tiny targets for engineering CHO cell phenotypes? Biotechnol Lett 33(1):11–21

Behrman S, Acosta-Alvear D, Walter P (2011) A CHOP-regulated microRNA controls rhodopsin expression. J Cell Biol 192(6):919–927

Bhaumik D et al (2009) MicroRNAs miR-146a/b negatively modulate the senescence-associated inflammatory mediators IL-6 and IL-8. Aging (Albany NY) 1(4):402–411

Bi JX, Shuttleworth J, Al-Rubeai M (2004) Uncoupling of cell growth and proliferation results in enhancement of productivity in p21CIP1-arrested CHO cells. Biotechnol Bioeng 85(7):741–749

Borth N et al (2005) Effect of increased expression of protein disulfide isomerase and heavy chain binding protein on antibody secretion in a recombinant CHO cell line. Biotechnol Prog 21(1):106–111
Calin GA et al (2002) Frequent deletions and down-regulation of micro-RNA genes miR15 and miR16 at 13q14 in chronic lymphocytic leukemia. Proc Natl Acad Sci U S A 99(24):15524–15529
Carvalhal AV, Marcelino I, Carrondo MJ (2003) Metabolic changes during cell growth inhibition by p27 overexpression. Appl Microbiol Biotechnol 63(2):164–173
Chan JA, Krichevsky AM, Kosik KS (2005) MicroRNA-21 is an antiapoptotic factor in human glioblastoma cells. Cancer Res 65(14):6029–6033
Chang TC et al (2008) Widespread microRNA repression by Myc contributes to tumorigenesis. Nat Genet 40(1):43–50
Chen JF et al (2006) The role of microRNA-1 and microRNA-133 in skeletal muscle proliferation and differentiation. Nat Genet 38(2):228–233
Chen K et al (2001) Engineering of a mammalian cell line for reduction of lactate formation and high monoclonal antibody production. Biotechnol Bioeng 72(1):55–61
Chen P, Harcum SW (2006) Effects of elevated ammonium on glycosylation gene expression in CHO cells. Metab Eng 8(2):123–132
Chen P, Harcum SW (2007) Differential display identifies genes in Chinese hamster ovary cells sensitive to elevated ammonium. Appl Biochem Biotechnol 141(2–3):349–359
Cheng AM et al (2005) Antisense inhibition of human miRNAs and indications for an involvement of miRNA in cell growth and apoptosis. Nucleic Acids Res 33(4):1290–1297
Chiang GG, Sisk WP (2005) Bcl-x(L) mediates increased production of humanized monoclonal antibodies in Chinese hamster ovary cells. Biotechnol Bioeng 91(7):779–792
Chung JY et al (2004) Effect of doxycycline-regulated calnexin and calreticulin expression on specific thrombopoietin productivity of recombinant Chinese hamster ovary cells. Biotechnol Bioeng 85(5):539–546
Cimmino A et al (2005) miR-15 and miR-16 induce apoptosis by targeting BCL2. Proc Natl Acad Sci U S A 102(39):13944–13949
Clarke C et al (2011) Predicting cell-specific productivity from CHO gene expression. J Biotechnol 151(2):159–165
Creighton CJ et al (2010) Molecular profiling uncovers a p53-associated role for microRNA-31 in inhibiting the proliferation of serous ovarian carcinomas and other cancers. Cancer Res 70(5):1906–1915
Dai R et al (2010) miR-221/222 suppression protects against endoplasmic reticulum stress-induced apoptosis via p27(Kip1)- and MEK/ERK-mediated cell cycle regulation. Biol Chem 391(7):791–801
de Boer L, Gray PP, Sunstrom NA (2004) Enhanced productivity of G1 phase Chinese hamster ovary cells using the GADD153 promoter. Biotechnol Lett 26(1):61–65
Dinnis DM, James DC (2005) Engineering mammalian cell factories for improved recombinant monoclonal antibody production: lessons from nature? Biotechnol Bioeng 91(2):180–189
Druz A et al (2011) A novel microRNA mmu-miR-466h affects apoptosis regulation in mammalian cells. Biotechnol Bioeng 108(7):1651–1661
Dunster CA, Cheeseman KH, Maddix SP (1997) The effect of oxidative stress on the production of the recombinant protein, interferon gamma, produced by Chinese hamster ovary cells in stirred-batch culture. Appl Microbiol Biotechnol 48(2):198–203
Eichner LJ et al (2010) miR-378(*) mediates metabolic shift in breast cancer cells via the PGC-1beta/ERRgamma transcriptional pathway. Cell Metab 12(4):352–361
El Ouaamari A et al (2008) miR-375 targets 3′-phosphoinositide-dependent protein kinase-1 and regulates glucose-induced biological responses in pancreatic beta-cells. Diabetes 57(10):2708–2717
Elias CB et al (2003) Improving glucose and glutamine metabolism of human HEK 293 and Trichoplusia ni insect cells engineered to express a cytosolic pyruvate carboxylase enzyme. Biotechnol Prog 19(1):90–97

Esau C et al (2004) MicroRNA-143 regulates adipocyte differentiation. J Biol Chem 279(50):52361–52365
Esau C et al (2006) miR-122 regulation of lipid metabolism revealed by in vivo antisense targeting. Cell Metab 3(2):87–98
Fasanaro P et al (2008) MicroRNA-210 modulates endothelial cell response to hypoxia and inhibits the receptor tyrosine kinase ligand Ephrin-A3. J Biol Chem 283(23):15878–15883
Fassnacht D et al (1999) Influence of bcl-2 on antibody productivity in high cell density perfusion cultures of hybridoma. Cytotechnology 30(1–3):95–106
Felli N et al (2005) MicroRNAs 221 and 222 inhibit normal erythropoiesis and erythroleukemic cell growth via kit receptor down-modulation. Proc Natl Acad Sci U S A 102(50):18081–18086
Figueroa B Jr et al (2007) Enhanced cell culture performance using inducible anti-apoptotic genes E1B-19K and Aven in the production of a monoclonal antibody with Chinese hamster ovary cells. Biotechnol Bioeng 97(4):877–892
Fox SR et al (2005) A detailed understanding of the enhanced hypothermic productivity of interferon-gamma by Chinese-hamster ovary cells. Biotechnol Appl Biochem 41(Pt 3):255–264
Fussenegger M et al (2000) Regulated overexpression of the survival factor bcl-2 in CHO cells increases viable cell density in batch culture and decreases DNA release in extended fixed-bed cultivation. Cytotechnology 32(1):45–61
Fussenegger M, Bailey JE (1998) Molecular regulation of cell-cycle progression and apoptosis in mammalian cells: implications for biotechnology. Biotechnol Prog 14(6):807–833
Gammell P (2007) MicroRNAs: recently discovered key regulators of proliferation and apoptosis in animal cells: Identification of miRNAs regulating growth and survival. Cytotechnology 53(1–3):55–63
Gao C et al (2010) Reduced microRNA-218 expression is associated with high nuclear factor kappa B activation in gastric cancer. Cancer 116(1):41–49
Gao P et al (2009) c-Myc suppression of miR-23a/b enhances mitochondrial glutaminase expression and glutamine metabolism. Nature 458(7239):762–765
Gauthier BR, Wollheim CB (2006) MicroRNAs: 'ribo-regulators' of glucose homeostasis. Nat Med 12(1):36–38
Hackl M et al (2011) Next-generation sequencing of the Chinese hamster ovary microRNA transcriptome: Identification, annotation and profiling of microRNAs as targets for cellular engineering. J Biotechnol 153(1–2):62–75
He L et al (2007) A microRNA component of the p53 tumour suppressor network. Nature 447(7148):1130–1134
Heal R, McGivan J (1998) Induction of calreticulin expression in response to amino acid deprivation in Chinese hamster ovary cells. Biochem J 329 (Pt 2):389–394
Hermeking H (2010) The miR-34 family in cancer and apoptosis. Cell Death Differ 17(2):193–199
Hua Z et al (2006) MiRNA-directed regulation of VEGF and other angiogenic factors under hypoxia. PLoS One 1:e116
Huang W et al (2010) Tonicity-responsive microRNAs contribute to the maximal induction of osmoregulatory transcription factor OREBP in response to high-NaCl hypertonicity. Nucleic Acids Res 39(2):475–485
Ifandi V, Al-Rubeai M (2005) Regulation of cell proliferation and apoptosis in CHO-K1 cells by the coexpression of c-Myc and Bcl-2. Biotechnol Prog 21(3):671–677
Imam JS et al (2010) MicroRNA-185 suppresses tumor growth and progression by targeting the Six1 oncogene in human cancers. Oncogene 29(35):4971–4979
Irani N, Beccaria AJ, Wagner R (2002) Expression of recombinant cytoplasmic yeast pyruvate carboxylase for the improvement of the production of human erythropoietin by recombinant BHK-21 cells. J Biotechnol 93(3):269–282
Jeong DW et al (2004) Modification of glycolysis affects cell sensitivity to apoptosis induced by oxidative stress and mediated by mitochondria. Biochem Biophys Res Commun 313(4):984–991
Johnson KC et al (2011) Conserved microRNAs in Chinese hamster ovary cell lines. Biotechnol Bioeng 108(2):475–480

Johnson SM et al (2005) RAS is regulated by the let-7 microRNA family. Cell 120(5):635–647
Kantardjieff A et al (2010) Transcriptome and proteome analysis of Chinese hamster ovary cells under low temperature and butyrate treatment. J Biotechnol 145(2):143–159
Kaufmann H et al (1999) Influence of low temperature on productivity, proteome and protein phosphorylation of CHO cells. Biotechnol Bioeng 63(5):573–582
Keane JT, Ryan D, Gray PP (2003) Effect of shear stress on expression of a recombinant protein by Chinese hamster ovary cells. Biotechnol Bioeng 81(2):211–220
Kefas B et al (2008) microRNA-7 inhibits the epidermal growth factor receptor and the Akt pathway and is down-regulated in glioblastoma. Cancer Res 68(10):3566–3572
Kim CH, Oh Y, Lee TH (1997) Codon optimization for high-level expression of human erythropoietin (EPO) in mammalian cells. Gene 199(1–2):293–301
Kim NS, Lee GM (2002) Response of recombinant Chinese hamster ovary cells to hyperosmotic pressure: effect of Bcl-2 overexpression. J Biotechnol 95(3):237–248
Kim SH, Lee GM (2007) Down-regulation of lactate dehydrogenase-A by siRNAs for reduced lactic acid formation of Chinese hamster ovary cells producing thrombopoietin. Appl Microbiol Biotechnol 74(1):152–159
Korke R et al (2004) Large scale gene expression profiling of metabolic shift of mammalian cells in culture. J Biotechnol 107(1):1–17
Kramer O, Klausing S, Noll T (2010) Methods in mammalian cell line engineering: from random mutagenesis to sequence-specific approaches. Appl Microbiol Biotechnol 88(2):425–436
Krutzfeldt J, Stoffel M (2006) MicroRNAs: a new class of regulatory genes affecting metabolism. Cell Metab 4(1):9–12
Ku SC et al (2008) Effects of overexpression of X-box binding protein 1 on recombinant protein production in Chinese hamster ovary and NS0 myeloma cells. Biotechnol Bioeng 99(1):155–164
Kulshreshtha R et al (2007) A microRNA signature of hypoxia. Mol Cell Biol 27(5):1859–1867
Kwon RJ et al (2006) Artificial transcription factors increase production of recombinant antibodies in Chinese hamster ovary cells. Biotechnol Lett 28(1):9–15
Lal A et al (2009) miR-24-mediated downregulation of H2AX suppresses DNA repair in terminally differentiated blood cells. Nat Struct Mol Biol 16(5):492–498
Lanceta J et al (2010) MicroRNA group disorganization in aging. Exp Gerontol 45(4):269–278
Lee HJ, Palkovits M, Young WS (2006) 3rd, miR-7b, a microRNA up-regulated in the hypothalamus after chronic hyperosmolar stimulation, inhibits Fos translation. Proc Natl Acad Sci U S A 103(42):15669–15674
Lee ST et al (2010) Let-7 microRNA inhibits the proliferation of human glioblastoma cells. J Neurooncol 102(1):19–24
Lee YS et al (2005) Depletion of human micro-RNA miR-125b reveals that it is critical for the proliferation of differentiated cells but not for the down-regulation of putative targets during differentiation. J Biol Chem 280(17):16635–16641
Lewis BP, Burge CB, Bartel DP (2005) Conserved seed pairing, often flanked by adenosines, indicates that thousands of human genes are microRNA targets. Cell 120(1):15–20
Li G et al (2011) Role of miR-204 in the regulation of apoptosis, endoplasmic reticulum stress response, and inflammation in human trabecular meshwork cells. Invest Ophthalmol Vis Sci 52(6):2999–3007
Li N et al (2011) Increased expression of miR-34a and miR-93 in rat liver during aging, and their impact on the expression of Mgst1 and Sirt1. Mech Ageing Dev 132(3):75–85
Lim SF et al (2006) RNAi suppression of Bax and Bak enhances viability in fed-batch cultures of CHO cells. Metab Eng 8(6):509–522
Lim Y et al (2010) Engineering mammalian cells in bioprocessing—current achievements and future perspectives. Biotechnol Appl Biochem 55(4):175–189
Liu CJ et al (2010) miR-31 ablates expression of the HIF regulatory factor FIH to activate the HIF pathway in head and neck carcinoma. Cancer Res 70(4):1635–1644
Lovis P, Gattesco S, Regazzi R (2008) Regulation of the expression of components of the exocytotic machinery of insulin-secreting cells by microRNAs. Biol Chem 389(3):305–312

Lu J et al (2005) MicroRNA expression profiles classify human cancers. Nature 435(7043):834–838
Maes OC et al (2008) Murine microRNAs implicated in liver functions and aging process. Mech Ageing Dev 129(9):534–541
Majors BS, Betenbaugh MJ, Chiang GG (2007) Links between metabolism and apoptosis in mammalian cells: applications for anti-apoptosis engineering. Metab Eng 9(4):317–326
Martello G et al (2010) A MicroRNA targeting dicer for metastasis control. Cell 141(7):1195–1207
Mastrangelo AJ et al (2000) Part II. Overexpression of bcl-2 family members enhances survival of mammalian cells in response to various culture insults. Biotechnol Bioeng 67(5):555–564
Masuda S et al (2000) A new biological strategy for high productivity of recombinant proteins in animal cells by the use of hypoxia-response enhancer. Biotechnol Bioeng 67(2):157–164
Meng F et al (2006) Involvement of human micro-RNA in growth and response to chemotherapy in human cholangiocarcinoma cell lines. Gastroenterology 130(7):2113–2129
Mercille S, Massie B (1999) Apoptosis-resistant E1B-19K-expressing NS/0 myeloma cells exhibit increased viability and chimeric antibody productivity under perfusion culture conditions. Biotechnol Bioeng 63(5):529–543
Mersey BD, Jin P, Danner DJ (2005) Human microRNA (miR29b) expression controls the amount of branched chain alpha-ketoacid dehydrogenase complex in a cell. Hum Mol Genet 14(22):3371–3377
Muller D, Katinger H, Grillari J (2008) MicroRNAs as targets for engineering of CHO cell factories. Trends Biotechnol 26(7):359–365
O'Donnell KA et al (2005) c-Myc-regulated microRNAs modulate E2F1 expression. Nature 435(7043):839–843
Ohya T et al (2008) Improved production of recombinant human antithrombin III in Chinese hamster ovary cells by ATF4 overexpression. Biotechnol Bioeng 100(2):317–324
Ovcharenko D et al (2007) Genome-scale microRNA and small interfering RNA screens identify small RNA modulators of TRAIL-induced apoptosis pathway. Cancer Res 67(22):10782–10788
Ozturk SS, Riley MR, Palsson BO (1992) Effects of ammonia and lactate on hybridoma growth, metabolism, and antibody production. Biotechnol Bioeng 39(4):418–431
Patrick DM et al (2010) Defective erythroid differentiation in miR-451 mutant mice mediated by 14–3-3zeta. Genes Dev 24(15):1614–1619
Peng RW et al (2010) Munc18b regulates core SNARE complex assembly and constitutive exocytosis by interacting with the N-peptide and the closed-conformation C-terminus of syntaxin 3. Biochem J 431(3):353–361
Peng RW, Abellan E, Fussenegger M (2011) Differential effect of exocytic SNAREs on the production of recombinant proteins in mammalian cells. Biotechnol Bioeng 108(3):611–620
Plaisance V et al (2006) MicroRNA-9 controls the expression of Granuphilin/Slp4 and the secretory response of insulin-producing cells. J Biol Chem 281(37):26932–26942
Poy MN, Spranger M, Stoffel M (2007) microRNAs and the regulation of glucose and lipid metabolism. Diabetes Obes Metab 9 Suppl 2:67–73
Qin X et al (2010) MicroRNA-19a mediates the suppressive effect of laminar flow on cyclin D1 expression in human umbilical vein endothelial cells. Proc Natl Acad Sci U S A 107(7):3240–3244
Saito Y et al (2006) Specific activation of microRNA-127 with downregulation of the proto-oncogene BCL6 by chromatin-modifying drugs in human cancer cells. Cancer Cell 9(6):435–443
Sangokoya C, Telen MJ, Chi JT (2010) microRNA miR-144 modulates oxidative stress tolerance and associates with anemia severity in sickle cell disease. Blood 116(20):4338–4348
Sauerwald TM, Betenbaugh MJ, Oyler GA (2002) Inhibiting apoptosis in mammalian cell culture using the caspase inhibitor XIAP and deletion mutants. Biotechnol Bioeng 77(6):704–716
Sauerwald TM et al (2006) Combining caspase and mitochondrial dysfunction inhibitors of apoptosis to limit cell death in mammalian cell cultures. Biotechnol Bioeng 94(2):362–372
Scherr M et al (2007) Lentivirus-mediated antagomir expression for specific inhibition of miRNA function. Nucleic Acids Res 35(22):e149

Senger RS, Karim MN (2003) Effect of shear stress on intrinsic CHO culture state and glycosylation of recombinant tissue-type plasminogen activator protein. Biotechnol Prog 19(4):1199–1209
Shimizu S et al (2010) The let-7 family of microRNAs inhibits Bcl-xL expression and potentiates sorafenib-induced apoptosis in human hepatocellular carcinoma. J Hepatol 52(5):698–704
Si ML et al (2007) miR-21-mediated tumor growth. Oncogene 26(19):2799–2803
Sun W et al (2010) microRNA: a master regulator of cellular processes for bioengineering systems. Annu Rev Biomed Eng 12:1–27
Sung YH et al (2007) Influence of co-down-regulation of caspase-3 and caspase-7 by siRNAs on sodium butyrate-induced apoptotic cell death of Chinese hamster ovary cells producing thrombopoietin. Metab Eng 9(5–6):452–464
Taguchi A et al (2008) Identification of hypoxia-inducible factor-1 alpha as a novel target for miR-17–92 microRNA cluster. Cancer Res 68(14):5540–5545
Tan HK et al (2008) Overexpression of cold-inducible RNA-binding protein increases interferon-gamma production in Chinese-hamster ovary cells. Biotechnol Appl Biochem 49(Pt 4):247–257
Tang X et al (2009) Identification of glucose-regulated miRNAs from pancreatic {beta} cells reveals a role for miR-30d in insulin transcription. RNA 15(2):287–293
Tey BT et al (2000) Influence of bcl-2 on cell death during the cultivation of a Chinese hamster ovary cell line expressing a chimeric antibody. Biotechnol Bioeng 68(1):31–43
Tigges M, Fussenegger M (2006) Xbp1-based engineering of secretory capacity enhances the productivity of Chinese hamster ovary cells. Metab Eng 8:264-272
Tseng CF et al (2004) Antioxidant role of human haptoglobin. Proteomics 4(8):2221–2228
Weber M et al (2010) MiR-21 is induced in endothelial cells by shear stress and modulates apoptosis and eNOS activity. Biochem Biophys Res Commun 393(4):643–648
Weber W, Fussenegger M (2007) Inducible product gene expression technology tailored to bioprocess engineering. Curr Opin Biotechnol 18(5):399–410
Wong DC et al (2006) Targeting early apoptotic genes in batch and fed-batch CHO cell cultures. Biotechnol Bioeng 95(3):350–361
Xu C et al (2007) The muscle-specific microRNAs miR-1 and miR-133 produce opposing effects on apoptosis by targeting HSP60, HSP70 and caspase-9 in cardiomyocytes. J Cell Sci 120(Pt 17):3045–3052
Xu X et al (2011) The genomic sequence of the Chinese hamster ovary (CHO)-K1 cell line. Nat Biotechnol 29(8):735–741
Yallop CA, Svendsen I (2001) The effects of G418 on the growth and metabolism of recombinant mammalian cell lines. Cytotechnology 35(2):101–114
Yang M, Butler M (2000) Effects of ammonia on CHO cell growth, erythropoietin production, and glycosylation. Biotechnol Bioeng 68(4):370–380
Yang W, Paschen W (2008) Conditional gene silencing in mammalian cells mediated by a stress-inducible promoter. Biochem Biophys Res Commun 365(3):521–527
Yoon SK, Song JY, Lee GM (2003) Effect of low culture temperature on specific productivity, transcription level, and heterogeneity of erythropoietin in Chinese hamster ovary cells. Biotechnol Bioeng 82(3):289–298
Yu F et al (2010) Mir-30 reduction maintains self-renewal and inhibits apoptosis in breast tumor-initiating cells. Oncogene 29(29):4194–4204
Zamboni N, Sauer U (2009) Novel biological insights through metabolomics and 13C-flux analysis. Curr Opin Microbiol 12(5):553–558
Zanghi JA, Fussenegger M, Bailey JE (1999) Serum protects protein-free competent Chinese hamster ovary cells against apoptosis induced by nutrient deprivation in batch culture. Biotechnol Bioeng 64(1):108–119

Chapter 6
MicroRNAs for Enhancement of CHO Cell Proliferation and Stability: Insights from Neuroblastoma Studies

Raymond L. Stallings

Abstract The large-scale production of protein demands cell lines that have high proliferative potential and minimal tendencies to undergo apoptosis or senescence. Efforts in biotechnology have emphasized improving the genetic characteristics of CHO cells to enhance protein production. Many of the features of cell lines that are desirable for biotechnology enterprises, however, are the subject of anti-cancer research. Anti-cancer therapy attempts to decrease cell proliferation rates and increase the potential for cell differentiation and apoptosis to occur. Thus, it is possible for biotechnology oriented research to gain considerable insight from therapeutic targets identified in cancer related studies. Mature microRNAs (miRNA) are 19–22 nt RNA sequences that negatively regulate gene expression at a post-transcriptional level which are emerging as potential therapeutic targets for neoplasia. MiRNAs are significantly involved with cell proliferation, apoptosis, differentiation, senescence, cell migration and invasion in the context of both normal developmental processes and in malignant diseases. They can both promote or retard all of these cellular phenotypes, dependent on cellular context. The purpose of this chapter is to review the phenotypic effects of miRNAs in cancer, using the childhood tumor neuroblastoma as a model, with a view towards understanding how manipulation of miRNAs in CHO cells might improve the phenotypic features of the lines for biotechnology purposes.

Keywords CHO · Neuroblastoma · MicroRNA · Apoptosis · Senescence · Ccell proliferation · DNA methylation · Differentiation

---

R. L. Stallings
Cancer Genetics, Royal College of Surgeons in Ireland, Dublin, Ireland
e-mail: rstallings@rcsi.ie

National Children's Research Centre, Our Lady's Children's Hospital, Dublin, Ireland

N. Barron (ed.), *MicroRNAs as Tools in Biopharmaceutical Production*,
DOI 10.1007/978-94-007-5128-6_6, © Springer Science+Business Media Dordrecht 2012

6.1 Introduction

The Chinese hamster ovary cell line (CHO), derived by Puck et al. in (1958) has been widely used in the field of somatic cell genetics to identify novel genes through mutation analysis. A plethora of drug resistant and auxotrophic recessive genetic variants could be isolated at high frequency from this cell line (Siciliano et al. 1978; Siminovitch 1976). CHO cells have a very rapid rate of proliferation and possess a karyotype where one haploid set of homologues have normal banding patterns and a second set that have undergone considerable rearrangements to form stable and well defined marker chromosomes. The chromosome rearrangements have led to hemizygosity for some genomic regions that has contributed to the ease at which some recessive mutations can be isolated (Deaven and Petersen 1973). Numerous studies were undertaken to map genes to the rearranged chromosomes of CHO cells to further investigate the molecular mechanisms yielding the high frequency of CHO genetic variants, and to identify regions of synteny with the human genome (Adair et al. 1983; Adair et al. 1984; Stallings et al. 1984a; Stallings et al. 1983; Stallings et al. 1982; Crawford et al. 1985; Stallings et al. 1984b).

CHO cells have become the cell line of choice for the biomanufacturing of proteins because of the ease at which they can be genetically manipulated and because of their outstanding growth characteristics. It is therefore not surprising that considerable research has been undertaken to further enhance the growth characteristics of CHO cells for biotechnology purposes (Clarke et al. 2011a; Clarke et al. 2011b), including the use of microRNAs (miRNA) to further manipulate the phenotypic features of CHO cells (Barron et al. 2011). MiRNAs are tiny 20–22 nt sequences processed from much larger sequences that negatively regulate gene expression post-transcriptionally (He and Hannon 2004). These miRNAs act by binding to regions of homology on the 3′ UTRs of specific mRNA sequences, causing either the degradation or translational inhibition of the mRNA sequences on the RNA induced silencing complex (RISC). MiRNAs play major roles in normal physiological processes, and their dysregulation is intimately associated with the development and progression of many forms of cancer (He et al. 2007b). This is particularly true of the pediatric cancer neuroblastoma (Stallings 2009; Stallings et al. 2011; Stallings et al. 2010), which has been the subject of research in my laboratory. MiRNAs can act in a positive oncogenic manner, promoting cell division by targeting tumor suppressor genes, or as tumor suppressors by promoting apoptosis, cell differentiation or senescence (summarized in Fig. 6.1). Often the phenotypic effects of miRNAs are cancer cell type dependent, having completely opposite effects in different cancers.

Although the identification of miRNA targets that decrease cancer cell viability are exactly the opposite traits from genetically engineering CHO cells for increased proliferation, the same target miRNAs could be of interest for improving phenotypic features such as resistance to apoptosis, shorter cell cycle times, etc., for biotechnology purposes. This chapter is to review some of the lessons learned about miRNAs in the cancer field, with particular emphasis on the relevance to problems in biotechnology.

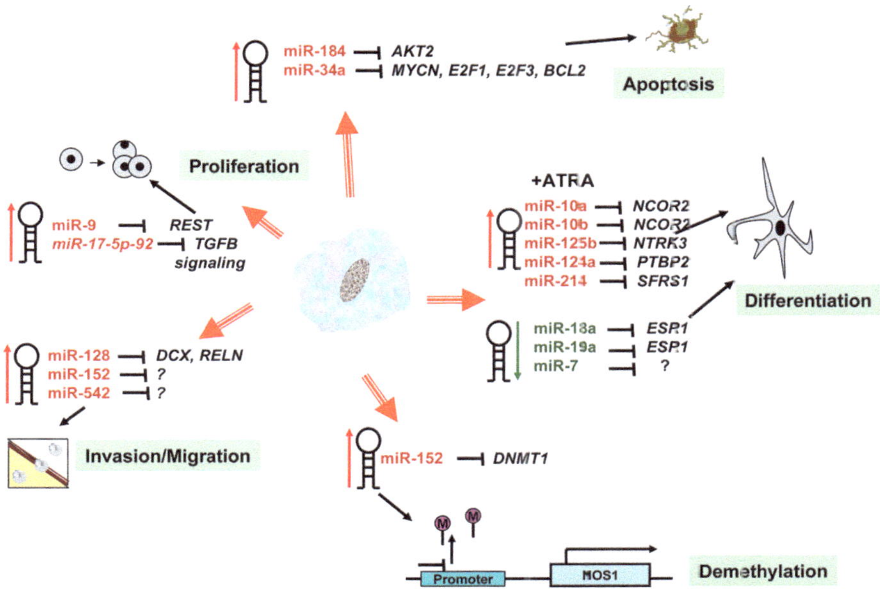

Fig. 6.1 Summary of miRNAs involved with neuroblastoma cell proliferation, apoptosis, differentiation and invasion, along with some of the validated mRNA targets

6.2 Oncogenic miRNAs in Neuroblastoma

One of the most important oncogenically acting miRNAs is the miR-17-5p-92 polycistronic cluster of six miRNAs mapping to human chromosome 13. The miR-17-5p-92 polycistronic cluster is over-expressed in many different forms of cancer, being directly regulated by the c-MYC oncogenic transcription factor (O'Donnell et al. 2005). The cluster is positively regulated in neuroblastoma by a related family member, MYCN, which is amplified and over-expressed in a subset of tumors that have a particularly aggressive phenotype. Members of the miR-17-5p-92 cluster are expressed at higher levels in neuroblastoma cell lines exhibiting over-expression of *MYCN* relative to cell lines with diploid copy number *MYCN* (Schulte et al. 2008). Using chromatin immunoprecipitation and luciferase reporter constructs, Fontana et al. (2008) demonstrated that MYCN protein directly binds to the promoter region of the miR-17-5p-92 cluster, resulting in up-regulation of the cluster. This group further demonstrated that ectopic over-expression of a single member of the cluster, miR-17-5p, increases proliferation rates of neuroblastoma cells both *in vitro* and *in vivo* (Fontana et al. 2008). The molecular mechanism of action for members of this polycistronic cluster is through targeting multiple members of the TGF-β canonical signaling pathway, including *TGFBR2*, *SMAD2*, *SMAD4* and *p21* (Fontana et al. 2008; Mestdagh et al. 2010a). Individual miRNAs within the cluster are conserved between humans and rodents, so that it would be of interest to determine if artificially increasing miR-17-5p-92 levels in CHO cells might further improve the characteristics of the line for bioprocessing purposes.

6.3 Apoptosis Promoting miRNAs

In general, more miRNAs appear to be down-regulated than up-regulated in cancers, indicating that they have major tumor suppressive effects (Bray et al. 2009). A number of miRNAs cause a variety of cancer cell types to undergo apoptosis when ectopically over-expressed, including both miR-34a and miR-184 in neuroblastoma. My research group was the first to demonstrate that miR-34a has tumor suppressor functions in cancer (Welch et al. 2007). MiR-34a maps to a region on human chromosome 1p which is frequently deleted in neuroblastoma tumors, resulting in lower expression of this miRNA. Ectopic over-expression of this miRNA in neuroblastoma cell lines leads to the arrest of cell proliferation and the induction of a caspase-mediated apoptotic pathway. miR-34a targets several genes with pro-proliferative or anti-apoptotic effects, including the *E2F3* and *MYCN* transcription factors (Cole et al. 2008; Wei et al. 2008; Welch et al. 2007), *BCL2* (Bommer et al. 2007), *CCND1* (Sun et al. 2008) and *CDK6* (Sun et al. 2008). Intriguingly, miR-34 family members are directly up-regulated by the tumor suppressive p53 transcription factor, which is mutated in greater than 50 % of all cancers (Bommer et al. 2007; Chang et al. 2007; He et al. 2007a; Raver-Shapira et al. 2007; Tarasov et al. 2007).

MiR-184 is another miRNA with pro-apoptotic effects in neuroblastoma, as originally demonstrated by Chen et al (Chen and Stallings 2007). Remarkably, even though miR-184 is predicted to target 100s of mRNA sequences, Foley et al. (2010) determined that the molecular mechanism leading to apoptosis was due solely to the targeting of the pro-survival serine-threonine kinase *AKT2*. siRNA-mediated knockdown of AKT2 by itself resulted in apoptosis, and an apoptotic phenotype caused by miR-184 targeting of *AKT2* could be rescued by co-transfection of miR-184 with an *AKT2* expression plasmid lacking a 3′ UTR target site.

MiR-34a has many conserved binding sites between vertebrate genomes, indicating that the engineering of CHO cells with reduced levels of miR-34a could be of interest. The miR-184/*AKT2* interaction site, on the other hand, is poorly conserved, and might not be of interest for further studies in CHO.

6.4 miRNAs Regulating Cell Differentiation

An important feature of neuroblastoma cells, and other cancer cell types, is their ability to undergo differentiation following treatment with retinoic acid or other differentiating agents (Stallings et al. 2011). Retinoic acid (RA) is a vitamin A derivative which activates gene transcription by binding to retinoic acid receptors (RAR) which are attached to retinoic acid response elements, which are short DNA sequence motifs mapping to primary RA response genes. RA causes a co-repressor complex, which includes the nuclear receptor co-repressor 2 (*NCOR2*) protein, to be displaced from the RAR heterodimer complex by a co-activator complex. It is now clear that a number of miRNAs are significantly up-regulated in response to retinoic acid treatment of neuroblastoma cell lines, indicating a potential role in the process of differentiation (Beveridge et al. 2009; Chen et al. 2010; Chen and Stallings 2007; Foley et al. 2011; Le et al. 2009; Meseguer et al. 2011; Ragusa et al. 2010). In particular, functional studies involving ectopic over-expression of miR-125b, miR-10a/b, and miR-214 indicate that these miRNAs promote neurite out-growth and the activation of markers of differentiation such as GAP43 and TUBB (Chen et al. 2010; Foley et al. 2011; Le et al. 2009). Conversely, inhibition of endogenous miR-7 and miR-18a promotes neuroblastoma cell differentiation *in vitro*. Remarkably, miR-10a/b targeting of the *NCOR2* mRNA, with a corresponding decrease in NCOR2 protein, is solely responsible for triggering the commencement of differentiation. It was demonstrated by Foley et al. (2011) that the NCOR2 co-repressor complex, in addition to being displaced by RA, must also decrease via a miRNA-mediated mechanism. The displacement and subsequent decrease in NCOR2 leads to the activation of primary retinoic acid responsive genes, which in turn launch a cascade of indirect effects that lead to a differentiated phenotype. The role of miRNAs in neuroblastoma cell differentiation is clearly very complex and a great amount of further investigation is required to elucidate these interesting and intricate regulatory networks.

6.5 miRNAs Involved with DNA Methylation

The hypermethylation of gene promoter regions plays a major role in inactivating tumor suppressor genes in the development of cancer (Kulis and Esteller 2010). Recently, it was also demonstrated that genome-wide DNA methylation alterations occur during the process of retinoic acid induced neuroblastoma differentiation, which appear to be mediated by miRNAs that target DNA methyltransferase (Das et al. 2010). In studies carried out by Das et al., a large set of genes were determined to be de-methylated in response to RA, consistent with the decrease in DNA methyltransferase activity. To determine a potential mechanism for the decrease in DNMT expression, a miRNA expression screen was performed pre- and post-RA which identified 17 up-regulated miRNAs. One of the up-regulated miRNAs, miR-152, was determined to directly target *DNMT1*, which maintains methylation patterns during cell division with a preference for hemi-methylated DNA (Hermann et al. 2004). Interestingly, other miRNAs up-regulated in response to RA, such as miR-125a/b and miR-26a/b, are predicted to target *DNMT3B*, and thus might also contribute to changes in the DNA methylome during differentiation.

An example of one of the genes that was demethylated and over expressed in several neuroblastoma cell lines treated with ATRA was nitric acid synthetase (*NOS1)* (Das et al. 2010), which catalyzes the generation of nitric oxide and has been implicated in neuroblastoma cell proliferation and differentiation (Ciani et al. 2004).

Ciani et al. (2004) reported that over expression of *NOS1* in SK-N-BE cells results in increased nitric oxide levels and decreased cell proliferation.

Thus, epigenetic regulation through DNA methylation appears to be controlled by miRNAs that directly inhibit DNA methyltransferases. Such miRNAs are clearly capable of launching a cascade of alterations to the cellular transcriptome through both direct and indirect effects. Alterations in DNA methylation certainly control the expression of numerous genes in CHO cells, and DNA methylation can contribute to the loss of productivity in the production of proteins in this cell line (Yang et al. 2010). The human miR-152/DNMT1 interaction site is broadly conserved in vertebrates, so it is possible that miRNAs could be mediating DNA methylation changes that negatively impact upon protein production in CHO cells through the targeting of DNA methytransferases.

6.6 miRNAs that Inhibit Cell Invasion and Migration Capacity

Ectopic over-expression of miR-152, which directly targets and inhibits DNA methyltransferase 1, as described above, also negatively impacts the neuroblastoma cell invasive potential without affecting rates of cell proliferation (Das et al. 2010). Other miRNAs, such as miR-128, also reduce neuroblastoma cell motility and invasiveness. For miR-128 this is accomplished by inhibiting the *Reelin* and *DCX* genes following ectopic over-expression (Evangelisti et al. 2009). In addition, miR-542-5p, when ectopically up-regulated, was determined to significantly inhibit neuroblastoma cell invasiveness without any change in the *in vitro* rate of cell proliferation (Bray et al. 2011). In the same study it was determined that ectopic over-expression of miR-542-5p inhibited tumor growth in an orthotopic murine xenograft model of the disease. Interestingly, this miRNA has no impact on *in vitro* cell proliferation rates. The mRNA targets of miR-542-5p that result in the decreased invasive potential and *in vivo* tumor growth are unknown.

6.7 miRNAs Targeting Larger Ultra Conserved Non-Coding RNAs

MiRNAs are only one type of non-coding RNA. In addition to miRNAs, there are literally thousands of longer length non-coding RNAs, the vast majority of which have unknown functions (Bertone et al. 2004). A subset of longer non-coding RNAs are transcribed from ultra-conserved regions (UCR) of the genome, which are defined as being 200 bp in length and having 100 % sequence conservation between human, mouse and rat genomes. Four hundred eighty-one UCRs have been identified, of which, $\sim$93 % are transcribed in one or more human tissues (i.e. T-UCRs) (Calin et al. 2007). Most significantly, approximately 9 % of T-UCRs are differentially expressed in cancer versus normal tissue, indicating a potential role in cancer (Calin et al. 2007). siRNA-mediated inhibition of one T-UCR in a colon cancer cell line was determined to have anti-proliferative effects (Calin et al. 2007). More recently, Scaruffi et al. (2009) and Mestdagh et al. (2010b) have reported that T-UCRs are differentially expressed in clinically favorable versus unfavorable neuroblastoma tumor subtypes, indicating a potential role for these non-coding sequences in the development of this disease. Interestingly, there is also preliminary evidence indicating that T-UCRs are regulated by miRNAs (Scaruffi et al. 2009; Calin et al. 2007), adding another layer of complexity to an already complex miRNA regulatory network. In some instances, larger non-coding RNAs could be acting as "sponges", i.e. endogenous competitors for miRNAs (Ebert and Sharp 2010).

Given that T-UCRs have very significant conservation between human and rodent genomes, by definition, and that some T-UCRs have anti-proliferative effects on human cancer cells, studies on these sequences in CHO cells would be of potential interest for further improving the phenotypic characteristics of this cell line for protein production.

6.8 The Mechanism of MicroRNA Action is not Completely Understood

Although it is now definitively established that miRNAs target the 3′ UTRs of large numbers of mRNA sequences, resulting in either degradation of the mRNA or translational inhibition, there are a number of reports that indicate miRNAs also have other mechanisms of action. For example, let-7 family members can apparently enhance the translational efficiency of target mRNA sequences in cells undergoing cell cycle arrest, while inhibiting translation in actively proliferating cells (Vasudevan et al. 2007). Let-7 is also an interesting family of miRNAs from the standpoint of processing of precursor sequences into biologically active mature miRNAs. The RNA binding proteins LIN28 and LIN28B can bind to Let-7 RNA precursors, blocking their bioprocessing by either Drosha or Dicer into mature sequences (Heo et al. 2009). Additionally, RNA binding proteins can bind to 3′ UTR sequences and protect mRNA sequences from miRNA targeting (Kedde et al. 2007). The let-7 family of miRNAs is highly conserved in diverse organisms, indicating that studies of this family in CHO cells is warranted, particularly in view of the fact that these miRNAs play major roles in cancer.

It is important to realize that miRNAs can interact with complementary sequences in the 5′ UTR and in the coding region of genes, in addition to 3′ UTR regions (Duursma et al. 2008; Orom et al. 2008), although a proteome-wide study by Baek et al indicated that miRNA interaction with 3′ UTRs had greater efficacy than the coding regions (Baek et al. 2008). This is likely explained by experiments performed by Gu et al. (2009), where the authors demonstrated that extending a coding region through the 3′ UTR of the miRNA target site by modification of the stop codon resulted in the loss of a miRNAs ability to translationally inhibit the mRNA sequence. This provides a biological reason for why most miRNA target sites are localized in

3′ UTRs. It is also of interest that some miRNAs can directly up-regulate gene transcription by direct interaction with regions of shared homology within the gene promoter region (Place et al. 2008).

All of the above discussion demonstrates that our understanding of the basic biology of miRNAs is far from complete. The elucidation of further unconventional modes of action of miRNAs is likely to have consequences for the use of miRNAs as therapeutics in cancer, as well as consequences for the use of CHO and other established cell lines as "protein factories" for biotechnology and industrial applications.

6.9 Concluding Remarks

Many of the miRNA/mRNA interaction sites that have been identified in cancer related studies are highly conserved between vertebrate genomes, indicating that the same miRNAs are also likely to be involved with the control of CHO cell viability. Artificial manipulation of miRNA levels in CHO cells holds considerable promise for increasing the bioprocessing potential of this cell line. The analysis of miRNAs in CHO cells, however, has been limited do to the lack of a complete DNA sequence map of either the normal Chinese hamster or CHO genomes. The recent completion of a DNA sequence map for the CHO genome should allow for the direct identification of miRNAs in CHO, opening up numerous additional avenues for biotechnology oriented research (Xu et al. 2011).

References

Adair GM, Stallings RL, Siciliano MJ (1984) Chromosomal rearrangements and gene expression in CHO cells: mapping of alleles for eight enzyme loci on CHO chromosomes Z3, Z4, Z5, and Z7. Somat Cell Mol Genet 10:283–295

Adair GM, Stallings RL, Nairn RS, Siciliano MJ (1983) High-frequency structural gene deletion as the basis for functional hemizygosity of the adenine phosphoribosyltransferase locus in Chinese hamster ovary cells. Proc Natl Acad Sci USA 80:5961–5964

Baek D, Villen J, Shin C, Camargo FD, Gygi SP, Bartel DP (2008) The impact of microRNAs on protein output. Nature 455:64–71

Barron N, Sanchez N, Kelly P, Clynes M (2011) MicroRNAs: tiny targets for engineering CHO cell phenotypes? Biotechnol Lett 33:11–21

Bertone P, Stolc V, Royce TE, Rozowsky JS, Urban AE, Zhu X, Rinn JL, Tongprasit W, Samanta M, Weissman S, Gerstein M, Snyder M (2004) Global identification of human transcribed sequences with genome tiling arrays. Science 306:2242–2246

Beveridge NJ, Tooney PA, Carroll AP, Tran N, Cairns MJ (2009) Down-regulation of miR-17 family expression in response to retinoic acid induced neuronal differentiation. Cell Signal 21:1837–1845

Bommer GT, Gerin I, Feng Y, Kaczorowski AJ, Kuick R, Love RE, Zhai Y, Giordano TJ, Qin ZS, Moore BB, Macdougald OA, Cho KR, Fearon ER (2007) p53-mediated activation of miRNA34 candidate tumor-suppressor genes. Curr Biol 17:1298–1307

Bray I, Bryan K, Prenter S, Buckley PG, Foley NH, Murphy DM, Alcock L, Mestdagh P, Vandesompele J, Speleman F, London WB, Mcgrady PW, Higgins DG, O'Meara A, O'Sullivan M, Stallings RL (2009) Widespread dysregulation of MiRNAs by MYCN amplification and chromosomal imbalances in neuroblastoma: association of miRNA expression with survival. PLoS One 4:e7850

Bray I, Tivnan A, Bryan K, Foley NH, Watters KM, Tracey L, Davidoff AM, Stallings RL (2011) MicroRNA-542–5p as a novel tumor suppressor in neuroblastoma. Cancer Lett 303:56–64

Calin GA, Liu CG, Ferracin M, Hyslop T, Spizzo R, Sevignani C, Fabbri M, Cimmino A, Lee EJ, Wojcik SE, Shimizu M, Tili E, Rossi S, Taccioli C, Pichiorri F, Liu X, Zupo S, Herlea V, Gramantieri L, Lanza G, Alder H, Rassenti L, Volinia S, Schmittgen TD, Kipps TJ, Negrini M, Croce CM (2007) Ultraconserved regions encoding ncRNAs are altered in human leukemias and carcinomas. Cancer Cell 12:215–229

Chang TC, Wentzel EA, Kent OA, Ramachandran K, Mullendore M, Lee KH, Feldmann G, Yamakuchi M, Ferlito M, Lowenstein CJ, Arking DE, Beer MA, Maitra A, Mendell JT (2007) Transactivation of miR-34a by p53 broadly influences gene expression and promotes apoptosis. Mol Cell 26:745–752

Chen H, Shalom-Feuerstein R, Riley J, Zhang SD, Tucci P, Agostini M, Aberdam D, Knight RA, Genchi G, Nicotera P, Melino G, Vasa-Nicotera M (2010) miR-7 and miR-214 are specifically expressed during neuroblastoma differentiation, cortical development and embryonic stem cells differentiation, and control neurite outgrowth in vitro. Biochem Biophys Res Commun 394:921–927

Chen Y, Stallings RL (2007) Differential patterns of microRNA expression in neuroblastoma are correlated with prognosis, differentiation, and apoptosis. Cancer Res 67 976–983

Ciani E, Severi S, Contestabile A, Bartesaghi R (2004) Nitric oxide negatively regulates proliferation and promotes neuronal differentiation through N-Myc downregulation. J Cell Sci 117:4727–4737

Clarke C, Doolan P, Barron N, Meleady P, O'Sullivan F, Gammell P, Melville M, Leonard M, Clynes M (2011a) Large scale microarray profiling and coexpression network analysis of CHO cells identifies transcriptional modules associated with growth and productivity. J Biotechnol 155:350–359

Clarke C, Doolan P, Barron N, Meleady P, O'Sullivan F, Gammell P, Melville M, Leonard M, Clynes M (2011b) Predicting cell-specific productivity from CHO gene expression. J Biotechnol 151:159–165

Cole KA, Attiyeh EF, Mosse YP, Laquaglia MJ, Diskin SJ, Brodeur GM, Maris JM (2008) A functional screen identifies miR-34a as a candidate neuroblastoma tumor suppressor gene. Mol Cancer Res 6:735–742

Crawford BD, Enger MD, Griffith BB, Griffith JK, Hanners JL, Longmire JL, Munk AC, Stallings RL, Tesmer JG, Walters RA et al (1985) Coordinate amplification of metallothionein I and II genes in cadmium-resistant Chinese hamster cells: implications for mechanisms regulating metallothionein gene expression. Mol Cell Biol 5:320–329

Das S, Foley N, Bryan K, Watters KM, Bray I, Murphy DM, Buckley PG, Stallings RL (2010) MicroRNA mediates DNA demethylation events triggered by retinoic acid during neuroblastoma cell differentiation. Cancer Res 70:7874–7881

Deaven LL, Petersen DF (1973) The chromosomes of CHO, an aneuploid Chinese hamster cell line: G-band, C-band, and autoradiographic analyses. Chromosoma 41:129–144

Duursma AM, Kedde M, Schrier M, Le Sage C, Agami R (2008) miR-148 targets human DNMT3b protein coding region. RNA 14:872–877

Ebert MS, Sharp PA (2010) Emerging roles for natural microRNA sponges. Curr Biol 20:R858–861

Evangelisti C, Florian MC, Massimi I, Dominici C, Giannini G, Galardi S, Bue M C, Massalini S, McDowell HP, Messi E, Gulino A, Farace MG, Ciafre SA (2009) MiR-128 up-regulation inhibits Reelin and DCX expression and reduces neuroblastoma cell motility and invasiveness. FASEB J 23:4276–4287

Foley NH, Bray I, Watters KM, Das S, Bryan K, Bernas T, Prehn JH, Stallings RL (2011) MicroRNAs 10a and 10b are potent inducers of neuroblastoma cell differentiation through targeting of nuclear receptor corepressor 2. Cell Death Differ 18:1089–1098

Foley NH, Bray IM, Tivnan A, Bryan K, Murphy DM, Buckley PG, Ryan J, O'Meara A, O'Sullivan M, Stallings RL (2010) MicroRNA-184 inhibits neuroblastoma cell survival through targeting the serine/threonine kinase AKT2. Mol Cancer 9:83

Fontana L, Fiori ME, Albini S, Cifaldi L, Giovinazzi S, Forloni M, Boldrini R, Donfrancesco A, Federici V, Giacomini P, Peschle C, Fruci D (2008) Antagomir-17-5p abolishes the growth of therapy-resistant neuroblastoma through p21 and BIM. PLoS One 3:e2236

Gu S, Jin L, Zhang F, Sarnow P, Kay MA (2009) Biological basis for restriction of microRNA targets to the 3′ untranslated region in mammalian mRNAs. Nat Struct Mol Biol 16:144–150

He L, Hannon GJ (2004) MicroRNAs: small RNAs with a big role in gene regulation. Nat Rev Genet 5:522–531

He L, He X, Lim LP, De Stanchina E, Xuan Z, Liang Y, Xue W, Zender L, Magnus J, Ridzon D, Jackson AL, Linsley PS, Chen C, Lowe SW, Cleary MA, Hannon GJ (2007a) A microRNA component of the p53 tumour suppressor network. Nature 447:1130–1134

He X, He L, Hannon GJ (2007b) The guardian's little helper: microRNAs in the p53 tumor suppressor network. Cancer Res 67:11099–11101

Heo I, Joo C, Kim YK, Ha M, Yoon MJ, Cho J, Yeom KH, Han J, Kim VN (2009) TUT4 in concert with Lin28 suppresses microRNA biogenesis through pre-microRNA uridylation. Cell 138:696–708

Hermann A, Goyal R, Jeltsch A (2004) The Dnmt1 DNA-(cytosine-C5)-methyltransferase methylates DNA processively with high preference for hemimethylated target sites. J Biol Chem 279:48350–48359

Kedde M, Strasser MJ, Boldajipour B, Oude Vrielink JA, Slanchev K, Le Sage C, Nagel R, Voorhoeve PM, Van Duijse J, Orom UA, Lund AH, Perrakis A, Raz E, Agami R (2007) RNA-binding protein Dnd1 inhibits microRNA access to target mRNA. Cell 131:1273–1286

Kulis M, Esteller M (2010) DNA methylation and cancer. Adv Genet 70:27–56

Le MT, Xie H, Zhou B, Chia PH, Rizk P, Um M, Udolph G, Yang H, Lim B, Lodish HF (2009) MicroRNA-125b promotes neuronal differentiation in human cells by repressing multiple targets. Mol Cell Biol 29:5290–5305

Meseguer S, Mudduluru G, Escamilla JM, Allgayer H, Barettino D (2011) Micro-RNAs-10a and -10b contribute to retinoic acid-induced differentiation of neuroblastoma cells and target the alternative splicing regulatory factor SFRS1 (SF2/ASF). J Biol Chem 286:4150–4164

Mestdagh P, Bostrom AK, Impens F, Fredlund E, Van Peer G, De Antonellis P, Von Stedingk K, Ghesquiere B, Schulte S, Dews M, Thomas-Tikhonenko A, Schulte JH, Zollo M, Schramm A, Gevaert K, Axelson H, Speleman F, Vandesompele J (2010a) The miR-17–92 microRNA cluster regulates multiple components of the TGF-beta pathway in neuroblastoma. Mol Cell 40:762–773

Mestdagh P, Fredlund E, Pattyn F, Rihani A, Van Maerken T, Vermeulen J, Kumps C, Menten B, De Preter K, Schramm A, Schulte J, Noguera R, Schleiermacher G, Janoueix-Lerosey I, Laureys G, Powel R, Nittner D, Marine JC, Ringner M, Speleman F, Vandesompele J (2010b) An integrative genomics screen uncovers ncRNA T-UCR functions in neuroblastoma tumours. Oncogene 29:3583–3592

O' Donnell KA, Wentzel EA, Zeller KI, Dang CV, Mendell JT (2005) c-Myc-regulated microRNAs modulate E2F1 expression. Nature 435:839–843

Orom UA, Nielsen FC, Lund AH (2008) MicroRNA-10a binds the 5′UTR of ribosomal protein mRNAs and enhances their translation. Mol Cell 30:460–471

Place RF, Li LC, Pookot D, Noonan EJ, Dahiya R (2008) MicroRNA-373 induces expression of genes with complementary promoter sequences. Proc Natl Acad Sci USA 105:1608–1613

Puck TT, Cieciura SJ, Robinson A (1958) Genetics of somatic mammalian cells. III. Long-term cultivation of euploid cells from human and animal subjects. J Exp Med 108:945–956

Ragusa M, Majorana A, Banelli B, Barbagallo D, Statello L, Casciano I, Guglielmino MR, Duro LR, Scalia M, Magro G, Di Pietro C, Romani M, Purrello M (2010) MIR152, MIR200B, and MIR338, human positional and functional neuroblastoma candidates, are involved in neuroblast differentiation and apoptosis. J Mol Med 88:1041–1053

Raver-Shapira N, Marciano E, Meiri E, Spector Y, Rosenfeld N, Moskovits N, Bentwich Z, Oren M (2007) Transcriptional activation of miR-34a contributes to p53-mediated apoptosis. Mol Cell 26:731–743

Scaruffi P, Stigliani S, Moretti S, Coco S, De Vecchi C, Valdora F, Garaventa A, Bonassi S, Tonini GP (2009) Transcribed-Ultra Conserved Region expression is associated with outcome in high-risk neuroblastoma. BMC Cancer 9:441

Schulte JH, Horn S, Otto T, Samans B, Heukamp LC, Eilers UC, Krause M, Astrahantseff K, Klein-Hitpass L, Buettner R, Schramm A, Christiansen H, Eilers M, Eggert A, Berwanger B (2008) MYCN regulates oncogenic MicroRNAs in neuroblastoma. Int J Cancer 122:699–704

Siciliano MJ, Siciliano J, Humphrey RM (1978) Electrophoretic shift mutants in Chinese hamster ovary cells: evidence for genetic diploidy. Proc Natl Acad Sci USA 75:1919–1923

Siminovitch L (1976) On the nature of hereditable variation in cultured somatic cells. Cell 7:1–11

Stallings RL (2009) MicroRNA involvement in the pathogenesis of neuroblastoma: potential for microRNA mediated therapeutics. Curr Pharm Des 15:456–462

Stallings RL, Adair GM, Lin JC, Siciliano MJ (1984a) Expression and regional assignment of Chinese hamster ESD and rRNA genes associated with translocations giving rise to chromosomes Z1 and Z6 in CHO cells. Cytogenet Cell Genet 38:132–137

Stallings RL, Adair GM, Siciliano J, Greenspan J, Siciliano MJ (1983) Genetic effects of chromosomal rearrangements in Chinese hamster ovary cells: expression and chromosomal assignment of TK, GALK, ACP1, ADA, and ITPA loci. Mol Cell Biol 3:1967–1974

Stallings RL, Foley NH, Bray IM, Das S, Buckley PG (2011) MicroRNA and DNA methylation alterations mediating retinoic acid induced neuroblastoma cell differentiation. Semin Cancer Biol 21:283–290

Stallings RL, Foley NH, Bryan K, Buckley PG, Bray I (2010) Therapeutic targeting of miRNAs in neuroblastoma. Expert Opin Ther Targets 14:951–962

Stallings RL, Munk AC, Longmire JL, Hildebrand CE, Crawford BD (1984b) Assignment of genes encoding metallothioneins I and II to Chinese hamster chromosome 3: evidence for the role of chromosome rearrangement in gene amplification. Mol Cell Biol 4:2932–2936

Stallings RL, Siciliano MJ, Adair GM, Humphrey RM (1982) Structural and functional hemi- and dizygous Chinese hamster chromosome 2 gene loci in CHO cells. Somatic Cell Genet 8:413–422

Sun F, Fu H, Liu Q, Tie Y, Zhu J, Xing R, Sun Z, Zheng X (2008) Downregulation of CCND1 and CDK6 by miR-34a induces cell cycle arrest. FEBS Lett 582:1564–1568

Tarasov V, Jung P, Verdoodt B, Lodygin D, Epanchintsev A, Menssen A, Meister G, Hermeking H (2007) Differential regulation of microRNAs by p53 revealed by massively parallel sequencing: miR-34a is a p53 target that induces apoptosis and G1-arrest. Cell Cycle 6:1586–1593

Vasudevan S, Tong Y, Steitz JA (2007) Switching from repression to activation: microRNAs can up-regulate translation. Science 318:1931–1934

Wei JS, Song YK, Durinck S, Chen QR, Cheuk AT, Tsang P, Zhang Q, Thiele CJ, Slack A, Shohet J, Khan J (2008) The MYCN oncogene is a direct target of miR-34a. Oncogene 27:5204–5213

Welch C, Chen Y, Stallings RL (2007) MicroRNA-34a functions as a potential tumor suppressor by inducing apoptosis in neuroblastoma cells. Oncogene 26:5017–5022

Xu X, Nagarajan H, Lewis NE, Pan S, Cai Z, Liu X, Chen W, Xie M, Wang W, Hammond S, Andersen MR, Neff N, Passarelli B, Koh W, Fan HC, Wang J, Gui Y, Lee KH, Betenbaugh MJ, Quake SR, Famili I, Palsson BO (2011) The genomic sequence of the Chinese hamster ovary (CHO)-K1 cell line. Nat Biotechnol 29:735–741

Yang Y, Mariati Chusainow J, Yap MG (2010) DNA methylation contributes to loss in productivity of monoclonal antibody-producing CHO cell lines. J Biotechnol 147:180–185

Chapter 7
MicroRNA Detection Methods for Mammalian Cell Lines and Their Applications in Therapeutic Protein Production

Nan Lin and Carol Kreader

Abstract Three major categories of detection methods that are applicable to study miRNA expression in Chinese Hamster Ovary (CHO) cells are reviewed in this chapter.

Keywords Chinese Hamster Ovary cells · Biotherapeutics · Cell engineering · microRNA · qRT-PCR · Microarrays · Next-generation sequencing

7.1 Overview of MicroRNA Detection in Mammalian Cells

MicroRNAs (miRNAs) are 21–25 nt single-stranded non-coding RNAs that play important roles in gene expression regulation (Bartel 2004). Research in miRNA is among one of the fastest growing fields in life sciences. In applied research, miRNAs are proposed to be valuable biomarkers and cell engineering tools in therapeutic protein production (Muller et al. 2008). Sensitive and accurate detection methods for miRNAs are the foundation for basic and applied research alike. In the past decade, researchers use quantitative reverse-transcription PCR (qRT-PCR), microarrays and other direct detection methods to profile and quantify miRNA expression in mammalian cells. In the past five years, direct and indirect miRNA detection technologies have developed even more rapidly with increased sensitivity and throughput in response to the growing research needs.

To this day, miRNA expression remains understudied in the biopharmaceutical context, particularly in Chinese Hamster Ovary (CHO) cells, despite the fact that CHO being the most commonly used host cell system for therapeutic protein production. The majority of known mature miRNAs are highly conserved in evolution (Ambros et al. 2003; Lagos-Quintana et al. 2001; Berezikov et al. 2005), which provides the rationale for application of existing miRNA detection methods in CHO

N. Lin (✉)
Cell Sciences and Development, SAFC/Sigma-Aldrich, 2909 Laclede Avenue, MO 63103, Saint Louis, USA
e-mail: Nan.lin@sial.com

C. Kreader
Research Biotechnology, Sigma-Aldrich, 2909 Laclede Avenue, MO 63103, Saint Louis, USA
e-mail: Carol.Kreader@sial.com

N. Barron (ed.), *MicroRNAs as Tools in Biopharmaceutical Production*,
DOI 10.1007/978-94-007-5128-6_7, © Springer Science+Business Media Dordrecht 2012

cells, regardless of the model organism (human, mouse or rat) used in the original development work. With the recent publication of the Whole Genome Sequencing Data for a CHO K1 cell line (Xu et al. 2011), researchers can expect a rapid increase in validated miRNA reserach tools for CHO cells and studies for miRNA expression in biopharmaceutical applications. This review aims to introduce the commonly used miRNA detection methods in mammalian cells to biopharmaceutical researchers, in order to aid discovery efforts and eventually tame these "tiny targets" for cell engineering (Barron et al. 2011b).

There are some unique challenges for miRNA detection (Git et al. 2010; Hammond 2006; Li and Ruan 2009; Yin et al. 2008). MicroRNAs are short—mature miRNAs are only the size of a typical qRT-PCR primer or probe. They have highly similar sequences within each miR family therefore detection accuracy often requires discriminating a single nucleotide. Proper reference gene selection and normalization for miRNA detection also constitute technical obstacles. Lack of easy purification, selective enrichment and amplification methods also makes it difficult to detect low abundant miRNA species. Researchers use different strategies to overcome the technical obstacles and have successfully developed methods for miRNA detection in mammalian cells. Among the most commonly practiced are quantitative reverse-transcriptase PCR (qRT-PCR), microarray and next-generation sequencing (NGS) based methods (Git et al. 2010), all of which have been applied to studies that are relevant to biopharmaceutical cell culture. The major findings of these studies are reviewed in the following sections. This review also introduces several novel miRNA detection methods that have not yet been reported for biopharmaceutical applications but can be adapted and validated to suit such purposes.

7.2 RNA Isolation for miRNA Expression Analysis

High quality RNA is essential for accurate miRNA detection and profiling. MicroRNA isolation from mammalian cells can be performed using total RNA isolation methods that preserve small RNAs (Mraz et al. 2009). Tri reagent based methods are a good choice (Rio et al. 2010). Users should verify small RNA preservation when using silica based RNA isolation kits that typically only preserve RNA species >200 nt. Some detection methods provide options for small RNA enrichment, which is of particular importance in next-generation sequencing based methods (Creighton et al. 2009). Storage and handling of miRNA samples is similar to that of total RNA.

7.3 Quantitative RT-PCR

Many researchers consider qRT-PCR the "gold standard" for miRNA detection (Table 7.1). The specificity of miRNA qRT-PCR methods is achieved by incorporating designs that are unique for miRNAs. The reverse transcription step typically utilizes a 5′ sequence tag in the primer for the first-strand synthesis (Wang 2009; Chen et al. 2005; Schmittgen et al. 2008). The sequence tag is then used for the reverse

Table 7.1 Examples of Commercially Available qRT-PCR Platforms for miRNA Detection

Manufacturer	RT method	qPCR detection method	Additional features	Normalization controls	Sensitivity of detection reported	Linear range of detection reported
Applied Biosystems	Gene-specific RT with pooling options	Taqman® dual-labeled	Stem-loop RT primer	Panel of miRNAs validated	10 copies of synthetic miRNA	7-log range
Exiqon	Oligo-dT primers with 5′-universal tag and 3′- degenerate anchor	SYBR® Green	LNA	Panel of miRNAs validated	10 copies of synthetic miRNA	7-log range
Qiagen	Oligo-dT primers with universal tag	SYBR® Green	None reported	Five snoRNAs	2000 copies of synthetic miRNA*	4-log range
Agilent	Adaptor primer with oligo-dT and universal tag	EvaGreen®	None reported	U6 snRNA	10 copies of synthetic miRNA	7-log range

* Reported by Hollingshead et al. (2007) http://www.genetics.pitt.edu/forms/flyers/miRNArealtimepcrevaluation.pdf
All information is based on manufacturer's websites unless otherwise specified

qPCR primer binding, in order to increase specificity of the qRT-PCR and lengthen the amplicon for detection. For example, Chen and colleagues use a stem-loop design for the miR-specific RT primers that allow detection and to distinguish between precursors and mature miRNAs (Chen et al. 2005; Schmittgen et al. 2008). Another approach is to include a 5′ universal tag in the oligo dT RT primer after adding a polyA tail to the mature miRNAs (Jacobsen et al. 2011), and the 5′-universal tag is used as a miR-specific reverse qPCR primer binding site.

It is not surprising that researchers selected qRT-PCR, one of the most well-established molecular biology methods, in early miRNA studies before CHO genome sequences were publicly available. Gammell and colleagues investigated the miRNA expression changes in CHO K1 cells induced by cold shock and published the first paper examining miRNA expression in CHO (Gammell et al. 2007). They used qRT-PCR as their method for differential expression analysis. The majority of subsequent studies published in CHO miRNA all used qRT-PCR for differential expression analyses, either as the primary method or as the confirmatory method following microarray experiments (Barron et al. 2011a; Druz et al. 2011; Lin et al. 2010).

While miR-specific RT provides great specificity, it limits the throughput and increases cost for miRNA qRT-PCR. Simply pooling miR-specific primers for universal RT may lead to interference between miRs and decrease detection sensitivity and specificity for the lower abundant miRs. The polyA addition followed by oligo dT priming method reviewed previously can be used to achieve universal RT reaction; specificity is achieved by strategies in qPCR such as incorporating LNAs in the primers (http://www.exiqon.com/microrna-real-time-pcr). This approach, naturally, can be used to design qRT panels. Another common strategy for universal RT is to design panels with limited number (up to 381) of miR qRT assays, then perform RT of these miRNAs after experimental validation of specificity and sensitivity (https://products.appliedbiosystems.com). A novel bioinformatics method partitions miRNA sequences and subsequently balances RT annealing temperature homogeneity, primer dimer formation and sequence cross-reactivity to achieve simultaneous qRT-PCR of 96 cancer-related miRNAs (Wang 2009).

Utilizing some of the techniques to increase throughput, researchers now routinely take "quantitative PCR array" approaches that retain the advantages of qRT-PCR, but utilize high-throughput 96- or 384-well multiplexed qRT assays that focus on various pathways (Chen et al. 2009). Among the most recently developed methods is Taqman Low Density Array (TLDA) (Zhang et al. 2010; Gokhale et al. 2010; Wang et al. 2011) that uses microfluidics cards to format qRT-PCR reactions with very small reaction volumes to increase throughput of qRT.

Mechanistic studies on miRNA biogenesis call for detection of miR precursors (pri-miRs and pre-miRs) that are longer in length (Lee et al. 2002). The detection of the precursor hairpin requires thermo stable RT, since denaturing the hairpin is essential for first-strand synthesis. Although the stable hairpin exists in both precursors, some researchers choose to amplify the larger pri-miRNA precursor due to increased sensitivity and linearity that they observed (Schmittgen et al. 2008).

Last but certainly not least, normalization has always been a point of debate since the discovery of miRNA, even more so in the less characterized CHO cells. Ideally a control RNA should have similar stability and size as the miRNAs analyzed in qRT-PCR (Chen et al. 2005; Wong et al. 2007), which makes small RNAs better controls than some traditional housekeeping genes. Koh and colleagues used small nucleolar RNA U6 as the reference gene. Gammel et al. (Gammell et al. 2007) used 5 S RNA as the housekeeping gene for their qRT-PCR studies. Several researchers selected miR-16 as the internal control miR based on its expression stability in various experimental conditions and tissues (Peltier and Latham 2008; Wong et al. 2007; Lin et al. 2010). A very recent publication by Druz et al. (Druz et al. 2011) used let-7c as the reference gene for qRT based on their microarray data in CHO cells. Increasing reports describe effective applications of algorithms such as Norm Finder for evaluating and selection of reference genes for miR qRT in various cell lines and organisms (Peltier and Latham 2008; Wotschofsky et al. 2011; Zhu et al. 2011; Chen et al. 2011; Brattelid et al. 2011; Schaefer et al. 2010; Carlsson et al. 2010; Latham 2010; Mortarino et al. 2010). The Vandesompele lab who has published extensively on qRT-PCR normalization proposed using the mean expression value of all miRNAs as the normalizer (Mestdagh et al. 2009). All of these normalization methods should be easily applied to CHO cells.

7.4 Microarrays

In the past decade, microarray-based methods have matured and become standard practice for transcriptional profiling. Continued discovery of new miRNAs has led to rapid expansion of miRBASE (Griffiths-Jones et al. 2006, 2008): v17 contains 16772 hairpin precursors representing 19724 mature miRNA products in 153 species. Humans alone are reported to have 1,492 miRNAs. As a result, array-based profiling becomes increasingly attractive to perform studies in which qRT-PCR does not provide sufficient throughput. Numerous types of commercially available and custom made microarray platforms have been developed and updated with the discovery of miRNAs. The small size of the miRNAs constitutes one of the technical challenges for probe design. Current oligonucleotide array platforms use similar principles and workflow of detection as DNA arrays: probe immobilization, target labeling and hybridization. For probe design, the oligonucleotide arrays incorporate modified nucleic acids such as Locked Nucleic Acids (LNAs) (Castoldi et al. 2007) or simply increase probe sizes in order to normalize melting temperature, and achieve high affinity and single-nucleotide discrimination (Li and Ruan 2009; Yin et al. 2008). An extended hairpin design in addition to the hybridization sequence has been successfully applied to increase specificity and stabilize probe-target interaction (Wang et al. 2007). Bort and colleagues used a custom LNA based microarray platform and reported successful differential expression analysis of miRNA expression in lag, exponential, stationary and decay phases of CHO K1 batch cultures (Bort et al. 2011).

While dual- or single-fluorescent probe labeling and hybridization on glass slide array platforms remains the most common, many researches developed their own custom oligo DNA arrays for miRNA profiling with various surface chemistry, printing technologies, probe design and labeling techniques (Liu et al. 2008; Git et al. 2010).

Git and colleagues comprehensively evaluated six commercially available array platforms for miRNA profiling, and concluded that Agilent, Illumina and Exiqon microarrays demonstrated least interarray variation, while Exiqon array stood out for good signal to noise ratio. Agilent, Illumina and Ambion arrays demonstrated the best correlation with qRT-PCR in this study. All of the cited array platforms may be suitable for CHO miRNA discovery work, either by using custom-designed probes against sequenced CHO miRNAs, or by using human/mouse/rat arrays for detecting highly conserved miRNAs in CHO. Although the Git study was not performed in CHO, it can be used as a good reference for selecting array platform for CHO based on the authors' comprehensiveness.

Nevertheless, the technical challenges for miRNA arrays remain. Besides similar inherent difficulties with qRT-PCR, RNA amplification to increase signal intensity is not yet proven to be applicable to miR arrays. Contrary to most linear amplification labeling protocols for DNA arrays (Ginsberg 2005), only miRNA precursors, not mature miRs, can be amplified through indirect labeling (Liu et al. 2004). It is critical to have a one to one ratio between the precursor and mature miRNA for accurate detection after amplification. However, miRNA precursors and mature miRNAs often deviate from one to one ratio due to post-transcriptional regulation in miRNA biogenesis (Lee et al. 2008; Thomson et al. 2006). Hence potential precursor contamination (Yin et al. 2008) may lead to amplification bias and skewed results. Without effective signal amplification, it can be difficult to achieve an acceptable signal to noise ratio and low background solely by optimization of washing temperature and duration, which remains the case for miRNA arrays.

A novel hybridization based technology, microfluidic μParaflo® chips that can overcome some of the signal to noise challenges discussed above, has come into play as an alternative to glass slide microarrays. The platform enables massive parallel synthesis of probes in picoliter-scale reaction chambers. The miRNA is dual-color labeled, and hybridization is conducted in these reaction chambers. The detection sensitivity and accuracy of this technology has attracted researchers to perform miRNA profiling in various species (Vorwerk et al. 2008; Zhu et al. 2007; Guo et al. 2009; Ding et al. 2009; Zhang et al. 2008). Druz and colleagues successfully employed custom μParaflo® chips that included 714 mouse and rat miRNA probes for the discovery of apoptosis regulatory miRNAs, and identified the pro-apototic miRNA, mmu-miR-466 h in CHO cells (Druz et al. 2011). After this pioneering work, μParaflo® chips may become a common tool for CHO miRNA profiling as the field grows.

7.5 Other Hybridization Based Methods

Bead-based hybridization for miRNA detection utilizes 5 μm polystyrene beads coated with oligonucleotide probes to specifically capture biotinylated miRNA. The hybridization is carried out in the bead suspension, in contrast to the glass array hybridization methods discussed in the previous section. The BeadArray method for miRNA detection is adapted from the cDNA-mediated annealing, selection, extension and ligation assay (DASL® assay) (Fan et al. 2004). The labeled beads are randomly assembled and distributed for fluorescent analysis on a 96-well plate Array Matrix (Chen et al. 2008). Alternatively, the captured miRNA is detected by streptavidin-phycoerythrin followed by flow cytometric analysis (Lu et al. 2005). The beads can be labelled with up to 100 colors, which allows for numerous combinations of miRNAs (Biscontin et al. 2010; Siegrist et al. 2009; Wang et al. 2011). Bead arrays are considered by some to have higher sensitivity, accuracy and flexibility but lower cost than glass arrays, but have not yet been reported to be used in CHO miRNA studies.

Northern blotting, a well-established RNA analysis method, has been successfully applied to both mature and precursor miRNA detection. Researchers incorporated LNA in their probe design and optimized washing conditions to increase sensitivity and they also successfully increased resolution by altering cross-linking or electrophoresis procedures (Koscianska et al. 2011; Kim et al. 2010, Pall and Hamilton 2008; Varallyay et al. 2008; Valoczi et al. 2004). Lusi and colleagues developed a novel label-free electrochemical detection method that incorporates an inosine-modified guanine free capture probe. They reported detection sensitivity to be 0.1 pmol (Lusi et al. 2009). These methods can also be applied to CHO miRNA research.

As mechanistic studies in CHO advance, the need for *in situ* detection of miRNAs will increase. *In situ* detection for miRNAs can elucidate questions such as differential miRNA expression in sub-cellular compartments or co-localization of miRNAs and their putative target mRNAs. Probe design and labeling again is the key to the success of this technique. Early approaches include using hapten-conjugated or radioactively labeled RNA oligonucleotide probes (Thompson et al. 2007) and LNA-enhanced probes (Kloosterman et al. 2006; Nelson et al. 2006; Pena et al. 2009). A recent publication reported single-cell miRNA quantitative detection and imaging using LNA probes (Lu and Tsourkas 2009), which can be useful for evaluating miRNA expression in a cell-based fashion in CHO cell populations, or for miR target co-localization studies.

7.6 Next-Generation Sequencing (NGS) for miRNA Profiling

In the past three to four years, NGS methods have flourished in genomics and transcriptomics research. Sample preparation protocols have been simplified, and costs greatly reduced, compared with when NGS was in its infancy. These methods allow for absolute quantification of target sequences with greater dynamic range and remarkably higher throughput per experiment than microarrays (Schuster 2008; Buermans et al. 2010). Some of the drawbacks of hybridization based technologies discussed earlier can be avoided using NGS. The read lengths (50–100 bp) for the two commonly used NGS platforms for small RNA sequencing (Illumina Genome Analyzer by Illumina and SOLiD by Applied Biosystems), albeit much shorter than Sanger sequencing, are well-suited for small RNAs and their precursors (Creighton et al. 2009). The procedures for miRNA profiling using NGS start with small RNA enrichment typically by PAGE followed by RNA quality check. Then small RNA libraries are constructed with 5′- and 3′- sequencing tags incorporated, and RT reactions are performed to convert the tagged RNA to cDNA. Lastly, the fraction with adequate size is isolated and sequenced using established methods such as "sequencing by synthesis" (SBS) isothermal bridge amplification (Illumina) or bead-based emulsion PCR (ABI SOLiD) (Tian et al. 2010). NGS is particularly valuable for miRNA discovery work due to the simultaneous availability of sequence and quantity (read counts).

Additionally, NGS is the only method that can effectively detect isomir sequences that represent variations of the mature miRNAs from the same precursor, an important new aspect in miRNA biogenesis (Morin et al. 2008).

The first NGS publication in CHO cells emerged in 2010 (Johnson et al. 2011). Johnson and colleagues reported sequences of 350 mature miRNA species from four CHO lines in various culture conditions. The NGS was performed using the Illumina platform from six cDNA libraries, and the annotations were based on homology with human, mouse and rat. The majority of annotations were made based on alignment with mature miRNAs in miRBASE 15.0. Differential expression studies between the cell lines or conditions were not reported in this publication. Nevertheless, this was the first report of NGS in CHO miRNA studies. Hackl and colleagues published the second study on NGS sequencing of CHO miRNA transcriptome less than a year later. They reported 387 annotated mature miRNAs from six CHO lines in various culture conditions (Hackl et al. 2011). While choosing Illumina as the NGS platform, they took a different approach in annotation by using the stem-loop for conserved miRNA identification. Sophisticated bioinformatics procedures were used to predict 122 novel miRNA. This work is the first to use NGS for differential expression of miRNAs in CHO in serum and serum-free culture conditions. A third and very recent publication successfully used the CHO whole genome sequencing data, in addition to using miRBASE sequences, to identify 190 conserved miRNA hairpin precursors (Hammond et al. 2011). The authors performed differential expression analyses as well, and reported more than 80 % (158) of miRNAs as significantly up- or down-regulated in at least one recombinant protein producing cell line comparing with the parental CHO K1.

The SoLiD platform, on the other hand, is also employed for NGS miRNA discovery and profiling for human and mouse (Schulte et al. 2010), but has not yet been reported in CHO cells. Tian and colleagues reported the potential bias in miRNA quantitative measurements that could be attributed to cDNA library preparation protocol differences. They observed in miRNA expression profiling from an unsequenced genome (sheep) higher variability in read lengths and in end sequence reads, also higher occurrence of end secondary structures when using the SoLiD method (Tian et al. 2010). Some researchers successfully adapted a SOLiD protocol to suit the SBS platform (Buermans et al. 2010).

In conclusion, NGS may require more validation against some of the more mature methods discussed in this review. Git and Tian both reported good correlation of the SBS method with qRT-PCR. It is still considered good practice to verify specific findings from high-throughput studies, NGS in particular, by qRT-PCR. The three major methods for detecting miRNA in CHO are summarized in Table 7.2. As with

Table 7.2 Summary of Three Major Methods for miRNA Detection

Method	Sensitivity of detection reported	Normalization controls	Technology maturity and applications in CHO	Major advantages	Major technical challenges
qRT-PCR	Requires the least amount of total RNA; highest sensitivity (can detect as low as 10 copies of synthetic miRNA)	Small RNA genes validated	"Gold standard"	Most validated technique; relatively easy to perform; least demanding in instrumentation	Low throughput
Microarrays	Requires most amount of total RNA due to labeling methods, replicates and controls required (often > 100 μg); lowest sensitivity and can miss low abundant miRNA species	Various in-array control probes; dye-swap technical replicates	Five years of reported application and validation in CHO	High throughput	Labor intensive labeling and hybridization steps; Low signal to noise ratio
NGS	Requires somewhat large amount of total RNA (1–10 μg)	Not reported	Very recent (<2 years of application and validation)	High throughput; sequence and quantitative data available simultaneously	Technique under validation and development

miRNA expression analysis in other cell lines and organisms, it is good practice to use one of two high throughput methods, microarrays or NGS, followed by verification of qRT-PCR when enthusiastic researchers start to swim further into the unknown waters of CHO miRNA expression. This workflow may increase the confidence in selecting the miRNAs discovered to proceed to cell engineering work.

Acknowledgements The authors would like to thank Dr. Niall Barron and Dr. Kevin J. Kayser for their helpful discussions during preparation of this chapter.

References

Ambros V, Bartel B, Bartel DP, Burge CB, Carrington JC, Chen X, Dreyfuss G, Eddy SR, Griffiths-Jones S, Marshall M, Matzke M, Ruvkun G, Tuschl T (2003) A uniform system for microRNA annotation. Rna 9:277–299

Barron N, Kumar N, Sanchez N, Doolan P, Clarke C, Meleady P, O'Sullivan F, Clynes M (2011a) Engineering CHO cell growth and recombinant protein productivity by overexpression of miR-7. J Biotechnol 151:204–211

Barron N, Sanchez N, Kelly P, Clynes M (2011b) MicroRNAs: tiny targets for engineering CHO cell phenotypes? Biotechnol Lett 33:11–21

Bartel DP (2004) MicroRNAs: Genomics, biogenesis, mechanism, and function. Cell 116:281–297

Berezikov E, Guryev V, Van De Belt J, Wienholds E, Plasterk RH, Cuppen E (2005) Phylogenetic shadowing and computational identification of human microRNA genes. Cell 120:21–24

Biscontin A, Casara S, Cagnin S, Tombolan L, Rosolen A, Lanfranchi G, De Pitta C (2010) New miRNA labeling method for bead-based quantification. BMC Mol Biol 11:44

Bort JA, Hackl M, Hoflmayer H, Jadhav V, Harreither E, Kumar N, Ernst W, Grillari J, Borth N (2011) Dynamic mRNA and miRNA profiling of CHO-K1 suspension cell cultures. Biotechnol J.

Brattelid T, Aarnes EK, Helgeland E, Guvaag S, Eichele H, Jonassen AK (2011) Normalization strategy is critical for the outcome of miRNA expression analyses in the rat heart. Physiol Genomics 43:604–610

Buermans HP, Ariyurek Y, Van Ommen G, Den Dunnen JT T, Hoen P A (2010) New methods for next generation sequencing based microRNA expression profiling. BMC Genomics 11:716

Carlsson J, Helenius G, Karlsson M, Lubovac Z, Andren O, Olsson B, Klinga-Levan K (2010) Validation of suitable endogenous control genes for expression studies of miRNA in prostate cancer tissues. Cancer Genet Cytogenet 202:71–75

Castoldi M, Benes V, Hentze MW, Muckenthaler MU (2007) miChip: a microarray platform for expression profiling of microRNAs based on locked nucleic acid (LNA) oligonucleotide capture probes. Methods 43:146–152

Chen C, Ridzon DA, Broomer AJ, Zhou Z, Lee DH, Nguyen JT, Barbisin M, Xu NL, Mahuvakar VR, Andersen MR, Lao KQ, Livak KJ, Guegler KJ (2005) Real-time quantification of microRNAs by stem-loop RT-PCR. Nucleic Acids Res 33:e179

Chen D, Pan X, Xiao P, Farwell MA, Zhang B (2011) Evaluation and identification of reliable reference genes for pharmacogenomics, toxicogenomics, and small RNA expression analysis. J Cell Physiol, 226:2469–2477

Chen J, Lozach J, Garcia EW, Barnes B, Luo S, Mikoulitch I, Zhou L, Schroth G, Fan JB (2008) Highly sensitive and specific microRNA expression profiling using BeadArray technology. Nucleic Acids Res 36:e87

Chen Y, Gelfond JA, Mcmanus LM, Shireman PK (2009) Reproducibility of quantitative RT-PCR array in miRNA expression profiling and comparison with microarray analysis. BMC Genomics 10:407

Creighton CJ, Reid JG, Gunaratne PH (2009) Expression profiling of microRNAs by deep sequencing. Brief Bioinform, 10:490–497

Ding D, Zhang L, Wang H, Liu Z, Zhang Z, Zheng Y (2009) Differential expression of miRNAs in response to salt stress in maize roots. Ann Bot 103:29–38

Druz A, Chu C, Majors B, Santuary R, Betenbaugh M, Shiloach J (2011) A novel microRNA mmu-miR-466h affects apoptosis regulation in mammalian cells. Biotechnol Bioeng 108:1651–1661

Fan JB, Yeakley JM, Bibikova M, Chudin E, Wickham E, Chen J, Doucet D, Rigault P, Zhang B, Shen R, McBride C, Li HR, Fu XD, Oliphant A, Barker DL, Chee MS (2004) A versatile assay for high-throughput gene expression profiling on universal array matrices. Genome Res 14:878–885

Gammell P, Barron N, Kumar N, Clynes M (2007) Initial identification of low temperature and culture stage induction of miRNA expression in suspension CHO-K1 cells. J Biotechnol 130:213–218

Ginsberg SD (2005) RNA amplification strategies for small sample populations. Methods, 37:229–237

Git A, Dvinge H, Salmon-Divon M, Osborne M, Kutter C, Hadfield J, Bertone P, Caldas C (2010) Systematic comparison of microarray profiling, real-time PCR, and next-generation sequencing technologies for measuring differential microRNA expression. Rna, 16:991–1006

Gokhale A, Kunder R, Goel A, Sarin R, Moiyadi A, Shenoy A, Mamidipally C, Noronha S, Kannan S, Shirsat NV (2010) Distinctive microRNA signature of medulloblastomas associated with the WNT signaling pathway. J Cancer Res Ther 6:521–529

Griffiths-Jones S, Grocock RJ, Van Dongen S, Bateman A, Enright AJ (2006) miRBase: microRNA sequences, targets and gene nomenclature. Nucleic Acids Res 34:D140–144

Griffiths-Jones S, Saini HK, Van Dongen S, Enright AJ (2008) miRBase: tools for microRNA genomics. Nucleic Acids Res 36:D154–158

Guo J, Miao Y, Xiao B, Huan R, Jiang Z, Meng D, Wang Y (2009) Differential expression of microRNA species in human gastric cancer versus non-tumorous tissues. J Gastroenterol Hepatol 24:652–657

Hackl M, Jakobi T, Blom J, Doppmeier D, Brinkrolf K, Szczepanowski R, Bernhart SH, Siederdissen CH, Bort JA, Wieser M, Kunert R, Jeffs S, Hofacker IL, Goesmann A, Puhler A, Borth N, Grillari J (2011) Next-generation sequencing of the Chinese hamster ovary microRNA transcriptome: Identification, annotation and profiling of microRNAs as targets for cellular engineering. J Biotechnol 153:62–75

Hammond S, Swanberg JC, Polson SW, Lee KH (2011) Profiling conserved MicroRNA expression in recombinant CHO cell lines using Illumina sequencing. Biotechnol Bioeng.

Hammond SM (2006) microRNA detection comes of age. Nat Methods 3:12–13

Hollingshead et al (2007) Evaluation of Methods for Extraction and Realtime PCR Measurement of micro-RNA. http://www.genetics.pitt.edu/forms/flyers/miRNArealtimepcrevaluation.pdf

Jacobsen N, Andreasen D, Mouritzen P (2011) Profiling microRNAs by real-time PCR. Methods Mol Biol 732:39–54

Johnson KC, Jacob NM, Nissom PM, Hackl M, Lee LH, Yap M, Hu WS (2011) Conserved microRNAs in Chinese hamster ovary cell lines. Biotechnol Bioeng 108:475–480

Kim SW, Li Z, Moore PS, Monaghan AP, Chang Y, Nichols M, John B (2010) A sensitive non-radioactive northern blot method to detect small RNAs. Nucleic Acids Res 38:e98

Kloosterman WP, Wienholds E, De Bruijn E, Kauppinen S, Plasterk RH (2006) In situ detection of miRNAs in animal embryos using LNA-modified oligonucleotide probes. Nat Methods 3:27–29

Koscianska E, Starega-Roslan J, Czubala K, Krzyzosiak WJ (2011) High-resolution northern blot for a reliable analysis of microRNAs and their precursors. ScientificWorldJournal 11:102–117

Lagos-Quintana M, Rauhut R, Lendeckel W, Tuschl T (2001) Identification of novel genes coding for small expressed RNAs. Science 294:853–858

Latham GJ (2010) Normalization of microRNA quantitative RT-PCR data in reduced scale experimental designs. Methods Mol Biol 667:19–31

Lee EJ, Baek M, Gusev Y, Brackett DJ, Nuovo GJ, Schmittgen TD (2008) Systematic evaluation of microRNA processing patterns in tissues, cell lines, and tumors. Rna 14:35–42

Lee Y, Jeon K, Lee JT, Kim S, Kim VN (2002) MicroRNA maturation: stepwise processing and subcellular localization. Embo J 21:4663–4670

Li W, Ruan K (2009) MicroRNA detection by microarray. Anal Bioanal Chem 394:1117–1124

Lin N, Davis A, Bahr S, Borgschulte T, Achtien K, Kayser K (2010) Profiling highly conserved microrna expression in recombinant IgG-producing and parental chinese hamster ovary cells. Biotechnol Prog.

Liu CG, Calin GA, Meloon B, Gamliel N, Sevignani C, Ferracin M, Dumitru CD, Shimizu M, Zupo S, Dono M, Alder H, Bullrich F, Negrini M, Croce CM (2004) An oligonucleotide microchip for genome-wide microRNA profiling in human and mouse tissues. Proc Natl Acad Sci USA 101:9740–9744

Liu CG, Calin GA, Volinia S, Croce CM (2008) MicroRNA expression profiling using microarrays. Nat Protoc 3:563–578

Lu J, Getz G, Miska EA, Alvarez-Saavedra E, Lamb J, Peck D, Sweet-Cordero A, Ebert, BL, Mak RH, Ferrando AA, Downing JR, Jacks T, Horvitz HR, Golub TR (2005) MicroRNA expression profiles classify human cancers. Nature 435:834–838

Lu J, Tsourkas A (2009) Imaging individual microRNAs in single mammalian cells in situ. Nucleic Acids Res 37:e100

Lusi EA, Passamano M, Guarascio P, Scarpa A, Schiavo L (2009) Innovative electrochemical approach for an early detection of microRNAs. Anal Chem 81:2819–2822

Mestdagh P, Van Vlierberghe P, De Weer A, Muth D, Westermann F, Speleman F, Vandesompele J (2009) A novel and universal method for microRNA RT-qPCR data normalization. Genome Biol 10:R64

Morin RD, O'Connor MD, Griffith M, Kuchenbauer F, Delaney A, Prabhu AL, Zhao Y, Mcdonald H, Zeng T, Hirst M, Eaves CJ, Marra MA (2008) Application of massively parallel sequencing to microRNA profiling and discovery in human embryonic stem cells. Genome Res 18:610–621

Mortarino M, Gioia G, Gelain ME, Albonico F, Roccabianca P, Ferri E, Comazzi S (2010) Identification of suitable endogenous controls and differentially expressed microRNAs in canine fresh-frozen and FFPE lymphoma samples. Leuk Res 34:1070–1077

Mraz M, Malinova K, Mayer J, Pospisilova S (2009) MicroRNA isolation and stability in stored RNA samples. Biochem Biophys Res Commun 390:1–4

Muller D, Katinger H, Grillari J (2008) MicroRNAs as targets for engineering of CHO cell factories. Trends Biotechnol 26:359–365

Nelson PT, Baldwin DA, Kloosterman WP, Kauppinen S, Plasterk RH, Mourelatos Z (2006) RAKE and LNA-ISH reveal microRNA expression and localization in archival human brain. Rna 12:187–191

Pall GS, Hamilton AJ (2008) Improved northern blot method for enhanced detection of small RNA. Nat Protoc 3:1077–1084

Peltier HJ, Latham GJ (2008) Normalization of microRNA expression levels in quantitative RT-PCR assays: identification of suitable reference RNA targets in normal and cancerous human solid tissues. Rna 14:844–852

Pena JT, Sohn-Lee C, Rouhanifard SH, Ludwig J, Hafner M, Mihailovic A, Lim C, Holoch D, Berninger P, Zavolan M, Tuschl T (2009) miRNA in situ hybridization in formaldehyde and EDC-fixed tissues. Nat Methods 6:139–1341

Rio DC, Ares M Jr., Hannon GJ, Nilsen TW (2010) Purification of RNA using TRIzol (TRI reagent). Cold Spring Harb Protoc 2010:pdb prot5439

Schaefer A, Jung M, Miller K, Lein M, Kristiansen G, Erbersdobler A, Jung K (2010) Suitable reference genes for relative quantification of miRNA expression in prostate cancer. Exp Mol Med 42:749–758

Schmittgen TD, Lee EJ, Jiang J, Sarkar A, Yang L, Elton TS, Chen C (2008) Real-time PCR quantification of precursor and mature microRNA. Methods 44:31–38

Schulte JH, Marschall T, Martin M, Rosenstiel P, Mestdagh P, Schlierf S, Thor T, Vandesompele J, Eggert A, Schreiber S, Rahmann S, Schramm A (2010) Deep sequencing reveals differential expression of microRNAs in favorable versus unfavorable neuroblastoma. Nucleic Acids Res 38:5919–5928

Schuster SC (2008) Next-generation sequencing transforms today's biology. Nat Methods 5:16–18

Siegrist F, Singer T, Certa U (2009) MicroRNA Expression Profiling by Bead Array Technology in Human Tumor Cell Lines Treated with Interferon-Alpha-2a. Biol Proced Online 11:113–129

Thompson RC, Deo M, Turner DL (2007) Analysis of microRNA expression by in situ hybridization with RNA oligonucleotide probes. Methods 43:153–161

Thomson JM, Newman M, Parker JS, Morin-Kensicki EM, Wright T, Hammond SM (2006) Extensive post-transcriptional regulation of microRNAs and its implications for cancer. Genes Dev 20:2202–2207

Tian G, Yin X, Luo H, Xu X, Bolund L, Zhang X, Gan SQ, Li N (2010) Sequencing bias: comparison of different protocols of microRNA library construction. BMC Biotechnol 10:64

Valoczi A, Hornyik C, Varga N, Burgyan J, Kauppinen S, Havelda Z (2004) Sensitive and specific detection of microRNAs by northern blot analysis using LNA-modified oligonucleotide probes. Nucleic Acids Res 32:e175

Varallyay E, Burgyan J, Havelda Z (2008) MicroRNA detection by northern blotting using locked nucleic acid probes. Nat Protoc 3:190–196

Vorwerk S, Ganter K, Cheng Y, Hoheisel J, Stahler PF, Beier M (2008) Microfluidic-based enzymatic on-chip labeling of miRNAs. N Biotechnol 25:142–199

Wang B, Howel P, Bruheim S, Ju J, Owen LB, Fodstad O, XiY (2011) Systematic evaluation of three microRNA profiling platforms: microarray, beads array, and quantitative real-time PCR array. PLoS One 6:e17167

Wang H, Ach RA, Curry B (2007) Direct and sensitive miRNA profiling from low-input total RNA. Rna 13:151–159

Wang X (2009) A PCR-based platform for microRNA expression profiling studies. Rna 15:716–723

Wong L, Lee K, Russell C, Chen C (2007) Endogenous Controls for Real-Time Quantitation of miRNA Using TaqMan MicroRNA Assays. Application Note Applied Biosystems.

Wotschofsky Z, Meyer HA, Jung M, Fendler A, Wagner I, Stephan C, Busch J, Erbersdobler A, Disch AC, Mollenkopf HJ, Jung K (2011) Reference genes for the relative quantification of microRNAs in renal cell carcinomas and their metastases. Anal Biochem 417:233–241

Xu X, Nagarajan H, Lewis NE, Pan S, Cai Z, Liu X, Chen W, Xie M, Wang W, Hammond S, Andersen MR, Neff N, Passarelli B, Koh W, Fan HC, Wang J, Gui Y, Lee KH, Betenbaugh MJ, Quake SR, Famili I, Palsson BO (2011) The genomic sequence of the Chinese hamster ovary (CHO)-K1 cell line. Nat Biotechnol 29:735–741

Yin JQ, Zhao RC, Morris KV (2008) Profiling microRNA expression with microarrays. Trends Biotechnol 26:70–76

Zhang L, Zhao W, Valdez JM, Creighton, CJ, Xin L (2010) Low-density Taqman miRNA array reveals miRNAs differentially expressed in prostatic stem cells and luminal cells. Prostate, 70:297–304

Zhang Z, Wei L, Zou X, Tao Y, Liu Z, Zheng Y (2008) Submergence-responsive MicroRNAs are potentially involved in the regulation of morphological and metabolic adaptations in maize root cells. Ann Bot 102:509–519

Zhu HT, Dong QZ, Wang G, Zhou HJ, Ren N, Jia HL, Ye QH, Qin LX (2011) Identification of Suitable Reference Genes for qRT-PCR Analysis of Circulating microRNAs in Hepatitis B Virus-Infected Patients. Mol Biotechnol.

Zhu Q, Hong A, Sheng N, Zhang X, Matejko A, Jun KY, Srivannavit O, Gulari E, Gao X, Zhou X (2007) microParaflo biochip for nucleic acid and protein analysis. Methods Mol Biol 382:287–312

Chapter 8
Manipulating Levels of Specific MicroRNAs in Mammalian Cells

Carol Kreader and Nan Lin

Abstract In this chapter, we discuss why microRNA (miRNA) manipulation is of interest to cell line engineering, and review the various approaches used to alter miRNA levels. We review the design and use of both synthetic and expressed miRNAs and miRNA inhibitors, as well as targeted insertion and deletion.

Keywords miRNA mimic · miRNA inhibitor · antimiR · Targeted integration · Zinc finger nuclease · ZFN

8.1 Why Manipulate MicroRNA Levels?

MicroRNAs are attractive targets for cell line engineering to enhance recombinant protein expression. Like transcription factors, a single miRNA may regulate expression from multiple genes in multiple pathways to elicit a given phenotype. The power of miRNAs is clearly illustrated by a recent report that exogenous expression of a single transcript encoding the miR-302/367 cluster can initiate reprogramming in mouse or human fibroblasts to generate pluripotent stem cells (Anokye-Danso et al. 2011), a process otherwise performed by expressing four separate proteins. Unlike transcription factors, however, miRNAs will not compete with the recombinant protein for translation. They are processed from longer transcripts by two RNase III enzymes, Drosha and Dicer, which are not known to play any role in messenger RNA (mRNA) processing. In addition, miRNAs regulate expression in association with the RNA induced silencing complex (RISC), which will only interfere with recombinant protein expression if the encoded transcript includes a miRNA target site. Finally, miRNAs are readily imitated. Mature miRNA mimics and inhibitors are short enough to manufacture by standard oligonucleotide synthesis for transient transfection into cells, and several hairpin precursors plus flanking regions easily fit into lentiviral and other expression vectors for stable expression (Zhou et al. 2008; Scherr et al. 2007).

C. Kreader (✉)
Research Biotechnology, Sigma-Aldrich, 2909 Laclede Avenue, Saint Louis, MO 63103, USA
e-mail: Carol.Kreader@sial.com

N. Lin
Cell Sciences and Development, SAFC/Sigma-Aldrich, 2909 Laclede Avenue, Saint Louis, MO 63103, USA
e-mail: Nan.lin@sial.com

N. Barron (ed.), *MicroRNAs as Tools in Biopharmaceutical Production*,
DOI 10.1007/978-94-007-5128-6_8,

8.2 Transient miRNA Manipulation with Synthetic Mimics and Inhibitors

Synthetic, double stranded RNA (dsRNA) oligonucleotides are often transfected into cells to imitate or mimic an endogenous miRNA. These dsRNA oligos resemble siRNAs, except that they are usually comprised of the natural mature miRNA paired with its partial complement instead of two fully complementary strands. The sequences of all known mature miRNAs and their hairpin precursors are freely available from miRBase (Kozomara and Griffiths-Jones 2011) and can be accessed at http://www.mirbase.org/. Although only two Chinese hamster miRNAs had been submitted to miRBase at the time this chapter was written (cgr-miR-21 and cgr-miR-7) (Barron et al. 2011; Gammell et al. 2007), miRBase 18, contained 1,157 mouse and 1,921 human mature miRNA sequences, many of which are conserved. Both the Borth (Hackl et al. 2011) and Lee (Hammond et al. 2012) groups have published sequencing results for conserved miRNAs expressed in several CHO cell lines, which can be cross-checked to verify sequence conservation for miRNAs of interest. Furthermore, the Borth group computationally identified gene loci and precursor miRNA sequences for several hundred conserved CHO miRNAs and deposited the sequences in miRBase for addition to future updates (Hackl et al. 2011).

As with siRNAs, various modifications to the dsRNA structure have been shown to improve efficacy and/or stability. Two nucleotide extensions on one or both 3′-ends, either the natural 2 nucleotides, 2 uridines, or 2 thymidines, have been reported to improve performance. An amino group at the 5′-end of the partially complementary (i.e., non-miRNA) strand may inhibit loading into the RISC complex in favor of the mature miRNA sequence, and 2′-O-methyl or other modified nucleotides may improve stability (Czauderna et al. 2003). Alternatively, miRNA mimics are commercially available from several sources. Each has their own proprietary miRNA mimic design and synthesis processes, and will custom design mimics for any miRNA sequence. These companies also sell libraries or panels of individual mimics for all known human or mouse miRNAs. As an alternative approach to profiling miRNAs in cell lines by microarray or sequencing, libraries or panels of miRNAs can be transfected into cells in a medium or high throughput format to screen for desired phenotypes and identify miRNA targets for cell line engineering.

Whereas some miRNAs may enhance recombinant protein production when upregulated, others may interfere with functions essential for good production, and therefore, need to be inhibited for biopharmaceutical applications. Synthetic miRNA inhibitors, also called antimiRs, are usually single-stranded RNA (ssRNA) antisense to the miRNA (i.e., the reverse complement sequence). Modified nucleotides, such as 2′-O-methyl and locked nucleic acids (LNAs), are often substituted for standard nucleotides to improve stability and targeting (Davis et al. 2006).

Synthetic miRNA inhibitors are commercially available from most of the same sources as mimics. In addition, as with miRNA mimics, libraries of miRNA inhibitors can be screened to identify miRNAs whose inhibition gives a desirable phenotype.

8.3 Stable Expression of miRNAs or Inhibitors

While synthetic miRNA mimics and inhibitors may be useful to verify or identify candidate miRNAs for up or down regulation in transient assays, stable expression is preferred for recombinant protein production. Even the most stable modified oligonucleotide will be diluted to an ineffective level as cells divide. Furthermore, it is neither practical nor regulatory compliant in production to re-transfect cells every few days with synthetic miRNAs or inhibitors. For stable expression, constructs encoding miRNAs or miRNA inhibitors may be inserted into and expressed from the genome. Both plasmid and viral vectors are used to deliver miRNA and antimiR expression constructs. To express miRNAs, the DNA sequence encoding the miRNA hairpin precursor along with 100–200 bases upstream and downstream have been cloned between an appropriate RNA polymerase II (Pol II) promoter and a 3′-polyadenylation sequence. However, since miRNAs can be regulated post-transcriptionally (Lee et al. 2008; Thomson et al. 2006), processing of a miRNA into its mature, active form may be inhibited depending on cell line and growth conditions. As an alternative to the natural precursor, artificial miRNAs are often used. To construct an artificial miRNA, the primary sequence of an expressed miRNA is used as a scaffold or backbone, with the stem sequence replaced by that for the miRNA of interest (Amendola et al. 2009). Such artificial miRNAs are processed like the parental miRNA used for the scaffold (Zeng et al. 2002). miRNA hairpin sequences have also been inserted into an artificial intron in green fluorescent protein (GFP) for expression (Amendola et al. 2009; Qiu et al. 2008). GFP can be used to report successful transfection, transcription, and processing, and fluorescing cells can be isolated by Fluorescence Activated Cell Sorting (FACS).

Most miRNA expression constructs are transcribed from Pol II promoters, analogous to their natural counterparts. Viral promoters such as cytomegalovirus (CMV) are used for high level expression, while the phosphoglycerate kinase (PGK) promoter is often used for more moderate expression or in cells that suppress or silence viral promoters. Even higher levels of miRNA may be expressed from an RNA polymerase III (Pol III) promoter such as U6 (Zhou et al. 2008). Use of an inducible promoter, so that expression of the miRNA can be turned on or off at an appropriate stage, such as when the cells reach a desired density for recombinant protein production, may also be beneficial. In fact, conditional expression may be essential if the miRNA's activity or its loss is detrimental to cell viability or growth. Tet-inducible promoters, which can be turned on or off by adding or removing doxycycline from the culture medium, are commonly used to restrict expression to a desired growth phase or time-frame (Gossen and Bujard 1992).

Pre-made lentiviral constructs for expressing human and mouse miRNAs are available from several commercial sources. In addition, vectors with many of the features mentioned above are commercially available for clone-your-own miRNA expression, and several companies offer custom-made miRNA expression constructs as a service.

Stable inhibition of miRNA activity can be achieved by expressing transcripts that contain miRNA binding sites, i.e., sequences complementary to the miRNA, to act as competitive inhibitors. Two main types have been reported—"sponges" and "tough decoys" (TuDs). Sponge constructs use multiple copies of a miRNA binding site to "soak up" or sequester miRNAs so that they are not available to regulate their natural targets (Care et al. 2007; Ebert et al. 2007; Gentner et al. 2009). miRNA binding sites are often placed in the 3′UTR of GFP or other reporter protein so that loss of signal indicates sponge function. Sponges have been used to inhibit miRNA activity in cultured cells, as well as in intact flies (Loya et al. 2009). TuDs, on the other hand, are a Pol III transcribed RNA with two opposing single-stranded miRNA binding sites sandwiched between a double-stranded stem and a stem-loop. Although TuDs have just two miRNA binding sites per molecule, they are transcribed by Pol III, the enzyme which drives expression of some of the most abundant RNAs. In addition, TuDs were designed for efficient nuclear export and their secondary structure was optimized for inhibition (Haraguchi et al. 2009). miRNA inhibition by TuDs has also been demonstrated in various cell types (Lu et al. 2011; Sakurai et al. 2011). Sponge and TuD structures with mismatches or additional bases inserted in the middle of the miRNA binding sites inhibit miRNAs more effectively than perfect complements, likely because these mismatched or bulged structures can not be cleaved by RISC. Both published (Haraguchi et al. 2009) and our own unpublished results suggest that TuDs may inhibit miRNA activity more effectively than sponges, likely owing to a higher expression by Pol III.

8.4 Targeted Gene Editing

In most published work to date, stably expressed miRNAs and miRNA inhibitors were integrated semi-randomly into the genome. Semi-random integration by lentiviral or plasmid delivered constructs can disrupt genes with desired functions or activate those detrimental to recombinant protein expression. In addition, regions surrounding the insertion site influence transgene expression, such that expression can vary or become silenced because of location. While integration at a specific site by homologous recombination (HR) is fairly efficient in mouse embryonic stem (ES) cells (Capecchi 2005), such targeted integration is inefficient in other species and cell lines.

Engineered zinc finger nucleases (ZFNs) that target the desired integration site can facilitate HR such that targeted integration is sufficiently efficient for cell line engineering. ZFNs are chimeric proteins, with multiple zinc finger protein domains fused to the non-specific nuclease domain of FokI restriction enzyme. The zinc finger domains bind specific DNA sequences and FokI, which acts as a dimer, cuts DNA between two ZFN pairs. In cells, DNA cleaved by the FokI nuclease is repaired by non-homologous end-joining (NHEJ) or by HR. Providing a donor construct with a gene of interest flanked by sequences homologous to those on either side of the cut site can facilitate HR and lead to insertion of the donor construct (Fig 8.1). This type of targeted integration has been widely demonstrated for protein expression (Carroll 2011). Appropriate "safe-haven" sites for targeted integration, that is, sites

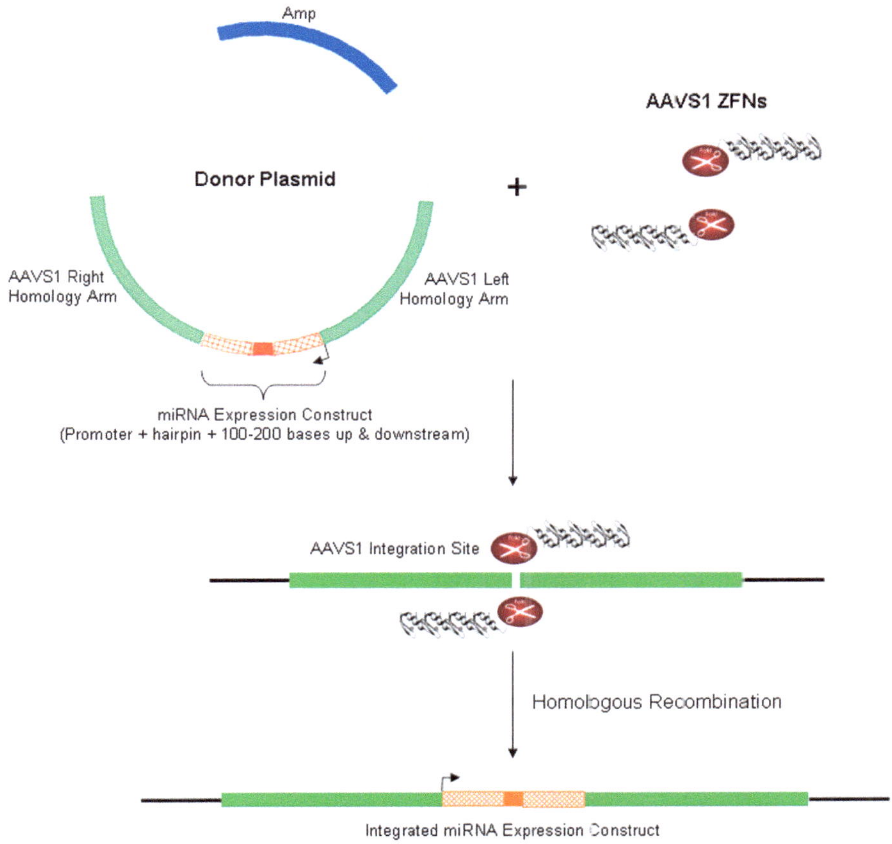

Fig. 8.1 ZFN-assisted targeted integration of a miRNA expression construct at the AAVS1 site. A miRNA hairpin/precursor sequence plus 100–200 nucleotides upstream and downstream are cloned after an appropriate promoter between two approximately 800 nucleotide sequences homologous to either side of the AAVS1 integration site to generate a donor plasmid. This plasmid, along with either a second plasmid or an RNA transcript that can express the two well-characterized ZFNs that target the AAVS1 integration site, are transfected into cells. After the AAVS1 ZFNs make a double-strand cut at the AAVS1 integration site, the miRNA expression construct can be integrated by homologous recombination

where transgenes can integrate without disrupting normal cellular functions or become silenced, have been identified in humans, mice, and rats. The adeno-associated virus integration site (AAVS1) in humans (DeKelver et al. 2010; Lombardo et al. 2011) and the Rosa26 locus in mice (Casola 2010) are well established sites for transgene integration and expression. Several labs have reported successful transgenic miRNA expression from Rosa26 in mice (Casola 2010; Medina et al. 2010), and our own unpublished results with ZFN-technology indicate that miRNAs and antimiRs can be targeted to and expressed from AAVS1 in human cells. Rats and hamsters also have the Rosa26 locus, but it remains to be determined whether this site can function as a safe-haven for transgene integration in these rodents as it does in mice.

Some miRNAs may interfere at all stages of cell growth and protein production to such an extent that complete and permanent knockout of the endogenous miRNA is a better approach than partial knockdown with an inhibitor. Many individual miRNAs have been knocked out in mice (Park et al. 2010) taking advantage of the well-developed genetic methods and the relatively high HR rate in this model organism (Capecchi 2005). In addition, a collection of miRNA targeting vectors to knockout 428 different mouse miRNAs are available from ES cell repositories (Prosser et al. 2011). However, miRNA knockout has not been reported in systems other than mouse where HR is less efficient. In theory, ZFNs that target a miRNA's hairpin precursor sequence could be used to disrupt its expression, but this has not been published.

8.5 Selecting miRNAs for Manipulation

Muller et al. (2008) proposed a number of miRNAs involved in growth, apoptosis, stress resistance, and translation in humans, mouse and/or rat, whose up or down-regulation could potentially enhance protein production in CHO cells. Other relevant miRNA targets can be identified by analyzing profile differences between cell lines, as recently reported by several groups of biopharmaceutical researchers (Barron et al. 2011; Hernández Bort 2012; Druz et al. 2011; Hackl et al. 2011; Johnson et al. 2011; Lin et al. 2011). Alternatively, libraries or panels of miRNA mimics and inhibitors can be screened for those that trigger a phenotype conducive to recombinant protein production, as mentioned above.

8.6 Use of miRNA Binding Sites

Binding sites for miRNAs that are highly expressed, either endogenously or exogenously, should be avoided in constructs used to express recombinant proteins. miRNA target sites can be predicted using online tools such as FindTar3 (http://bio.sz.tsinghua.edu.cn/).

Alternatively, a miRNA binding site can be included in constructs to prevent recombinant protein expression during growth stages in which the miRNA is highly expressed, or restrict expression to stages in which the miRNA is reduced or absent. Brown et al. demonstrated proof of concept for this approach in human dendritic cells using a GFP expression vector with four miR-155 target sites. GFP expression was high in immature dendritic cells while miR-155 expression was low, but was reduced 10-fold as cells matured and miR-155 expression increased 1,000-fold. (Brown et al. 2007). Additional examples of the use of miRNA binding sites to restrict transgene expression are reviewed in Brown and Naldini (2009).

8.7 Conclusions

miRNA manipulation is a viable strategy for cell line engineering. miRNAs can regulate multiple targets within a pathway, and are readily imitated due to their small size. As discussed above, synthetic miRNA mimics and inhibitors can be used to test predicted functions, as well as to screen for desired functions. Stable expression of miRNAs or miRNA inhibitors, especially from an inducible construct and with targeted integration into a "safe haven", can be used to perpetuate the phenotype for recombinant protein production. Finally, miRNA binding sites may be useful to restrict transgene expression temporally.

References

Amendola M, Passerini L, Pucci F, Gentner B, Bacchetta R, Naldini L (2009) Regulated and multiple miRNA and siRNA delivery into primary cells by a lentiviral platform. Mol Ther 17:1039–1052

Anokye-Danso F, Trivedi CM, Juhr D, Gupta M, Cui Z, Tian Y, Zhang Y, Yang W, Gruber PJ, Epstein JA, Morrisey EE (2011) Highly efficient miRNA-mediated reprogramming of mouse and human somatic cells to pluripotency. Cell Stem Cell 8:376–388

Barron N, Kumar N, Sanchez N, Doolan P, Clarke C, Meleady P, O'sullivan F, Clynes M (2011) Engineering CHO cell growth and recombinant protein productivity by overexpression of miR-7. J Biotechnol 151:204–211

Brown BD, Naldini L (2009) Exploiting and antagonizing microRNA regulation for therapeutic and experimental applications. Nat Rev Genetic 10:578–585

Brown BD, Gentner B, Cantore A, Colleoni S, Amendola M, Zingale A, Baccarini A, Lazzari G, Galli C, Naldini L (2007) Endogenous microRNA can be broadly exploited to regulate transgene expression according to tissue, lineage and differentiation state. Nat Biotechnol 25:1457–1466

Capecchi M (2005) Gene targeting in mice: functional analysis of the mammalian genome for the twenty-first century. Nat Rev Genet 6:507–512

Care A, Catalucci D, Felicetti F, Bonci D, Addario A, Gallo P, Bang ML, Segnalini P, Gu Y, Dalton ND, Elia L, Latronico MV, Hoydal M, Autore C, Russo MA, Dorn GW, 2nd, Ellingsen O, Ruiz-Lozano P, Peterson KL, Croce CM, Peschle C, Condorelli G (2007) MicroRNA-133 controls cardiac hypertrophy. Nat Med 13:613–618

Carroll D (2011) Genome engineering with zinc-finger nucleases. Genetics 188:773–782

Casola S (2010) Mouse models for miRNA expression: the ROSA26 locus. Methods Mol Biol 667:145–163

Czauderna F, Fechtner M, Dames S, Aygun H, Klippel A, Pronk GJ, Giese K, Kaufmann J (2003) Structural variations and stabilising modifications of synthetic siRNAs in mammalian cells. Nucleic Acids Res 31:2705–2716

Davis S, Lollo B, Freier S, Esau C (2006) Improved targeting of miRNA with antisense oligonucleotides. Nucleic Acids Res 34:2294–2304

Dekelver RC, Choi VM, Moehle EA, Paschon DE, Hockemeyer D, Meijsing SH, Sancak Y, Cui X, Steine EJ, Miller JC, Tam P, Bartsevich VV, Meng X, Rupniewski I, Gopalan SM, Sun HC, Pitz KJ, Rock JM, Zhang L, Davis GD, Rebar EJ, Cheeseman IM, Yamamoto KR, Sabatini DM, Jaenisch R, Gregory PD, Urnov FD (2010) Functional genomics, proteomics, and regulatory DNA analysis in isogenic settings using zinc finger nuclease-driven transgenesis into a safe harbor locus in the human genome. Genome Res 20:1133–1142

Druz A, Chu C, Majors B, Santuary R, Betenbaugh M, Shiloach J (2011) A novel microRNA mmu-miR-466h affects apoptosis regulation in mammalian cells. Biotechnol Bioeng 108:1651–1661

Ebert MS, Neilson JR, Sharp PA (2007) MicroRNA sponges: competitive inhibitors of small RNAs in mammalian cells. Nat Methods 4:721–726

Gammell P, Barron N, Kumar N, Clynes M (2007) Initial identification of low temperature and culture stage induction of miRNA expression in suspension CHO-K1 cells. J Biotechnol 130:213–218

Gentner B, Schira G, Giustacchini A, Amendola M, Brown BD, Ponzoni M, Naldini L (2009) Stable knockdown of microRNA in vivo by lentiviral vectors. Nat Methods 6:63–66

Gossen M, Bujard H (1992) Tight control of gene expression in mammalian cells by tetracycline-responsive promoters. Proc Natl Acad Sci USA 89:5547–5551

Hackl M, Jakobi T, Blom J, Doppmeier D, Brinkrolf K, Szczepanowski R, Bernhart SH, Siederdissen CH, Bort JA, Wieser M, Kunert R, Jeffs S, Hofacker IL, Goesmann A, Puhler A, Borth N, Grillari J (2011) Next-generation sequencing of the Chinese hamster ovary microRNA transcriptome: Identification, annotation and profiling of microRNAs as targets for cellular engineering. J Biotechnol 153:62–75

Hammond S, Swanberg JC, Polson SW, Lee KH (2012) Profiling conserved MicroRNA expression in recombinant CHO cell lines using Illumina sequencing. Biotechnol Bioeng 109:1371–1375

Haraguchi T, Ozaki Y, Iba H (2009) Vectors expressing efficient RNA decoys achieve the long-term suppression of specific microRNA activity in mammalian cells. Nucleic Acids Res 37:e43

Hernández Bort JA, Hackl M, Höflmayer H, Jadhav V, Harreither E, Kumar N, Ernst W, Grillari J, Borth N (2012) Dynamic mRNA and miRNA profiling of CHO-K1 suspension cell cultures. Biotechnol J 7:500–515

Johnson KC, Jacob NM, Nissom PM, Hackl M, Lee LH, Yap M, Hu WS (2011) Conserved microRNAs in Chinese hamster ovary cell lines. Biotechnol Bioeng 108:475–480

Kozomara A, Griffiths-Jones S (2011) miRBase: integrating microRNA annotation and deep-sequencing data. Nucleic Acids Res 39:D152–157

Lee EJ, Baek M, Gusev Y, Brackett DJ, Nuovo GJ, Schmittgen TD (2008) Systematic evaluation of microRNA processing patterns in tissues, cell lines, and tumors. Rna 14:35–42

Lin N, Davis A, Bahr S, Borgschulte T, Achtien K, Kayser K (2011) Profiling highly conserved microrna expression in recombinant IgG-producing and parental chinese hamster ovary cells. Biotechnol Prog 27:1163–1171

Lombardo A, Cesana D, Genovese P, Di Stefano B, Provasi E, Colombo DF, Neri M, Magnani Z, Cantore A, Lo Riso P, Damo M, Pello OM, Holmes MC, Gregory PD, Gritti A, Broccoli V, Bonini C, Naldini L (2011) Site-specific integration and tailoring of cassette design for sustainable gene transfer. Nat Methods 8:861–869

Loya CM, Lu CS, Van Vactor D, Fulga TA (2009) Transgenic microRNA inhibition with spatiotemporal specificity in intact organisms. Nat Methods 6:897–903

Lu Z, Li Y, Takwi A, Li B, Zhang J, Conklin DJ, Young KH, Martin R (2011) miR-301a as an NF-kappaB activator in pancreatic cancer cells. Embo J 30:57–67

Medina PP, Nolde M, Slack FJ (2010) OncomiR addiction in an in vivo model of microRNA-21-induced pre-B-cell lymphoma. Nature 467:86–90

Muller D, Katinger H, Grillari J (2008) MicroRNAs as targets for engineering of CHO cell factories. Trends Biotechnol 26:359–365

Park CY, Choi YS, McManus MT (2010) Analysis of microRNA knockouts in mice. Hum Mol Genet 19:R169–R175

Prosser HM, Koike-Yusa H, Cooper JD, Law FC, Bradley (2011) A resource of vectors and ES cells for targeted deletion of microRNAs in mice. Nat Biotechnol 29:840–845

Qiu L, Wang H, Xia X, Zhou H, Xu Z (2008) A construct with fluorescent indicators for conditional expression of miRNA. BMC Biotechnol 8:77

Sakurai K, Furukawa C, Haraguchi T, Inada K, Shiogama K, Tagawa T, Fujita S, Ueno Y, Ogata A, Ito M, Tsutsumi Y, Iba H (2011) MicroRNAs miR-199a-5p and -3p target the Brm subunit of SWI/SNF to generate a double-negative feedback loop in a variety of human cancers. Cancer Res 71:1680–1689

Scherr M, Venturini L, Battmer K, Schaller-Schoenitz M, Schaefer D, Dallmann I, Ganser A, Eder M (2007) Lentivirus-mediated antagomir expression for specific inhibition of miRNA function. Nucleic Acids Res 35:e149

Thomson JM, Newman M, Parker JS, Morin-Kensicki EM, Wright T, Hammond SM (2006) Extensive post-transcriptional regulation of microRNAs and its implications for cancer. Genes Dev 20:2202–2207

Zeng Y, Wagner EJ, Cullen BR (2002) Both natural and designed micro RNAs can inhibit the expression of cognate mRNAs when expressed in human cells. Mol Cell 9:1327–1333

Zhou H, Huang C, Xia XG (2008) A tightly regulated Pol III promoter for synthesis of miRNA genes in tandem. Biochim Biophys Acta 1779:773–779

Index

N. Barron (ed.), *MicroRNAs as Tools in Biopharmaceutical Production*,
DOI 10.1007/978-94-007-5128-6, © Springer Science+Business Media Dordrecht 2012

MIX
Papier aus verantwortungsvollen Quellen
Paper from responsible sources
FSC® C105338

If you have any concerns about our products,
you can contact us on
ProductSafety@springernature.com

In case Publisher is established outside the EU,
the EU authorized representative is:
Springer Nature Customer Service Center GmbH
Europaplatz 3, 69115 Heidelberg, Germany

Printed by Libri Plureos GmbH
in Hamburg, Germany